CHARLES VANCE

Emergency Communications: The Vital Role of Amateur Radio

First published by Mountaintop Publishing 2023

Powered by ChatGPT

First edition

This book was professionally typeset on Reedsy.
Find out more at reedsy.com

Contents

Preface

The world is full of uncertainties and emergencies can strike at any time. When disaster strikes, effective communication can play a crucial role in saving lives and minimizing the impact of the disaster. In such situations, ham radio operators have long been recognized for their role in providing reliable communication when other means of communication fail or become overloaded.

This book is a comprehensive guide to emergency communications and the vital role that ham radio operators play in crisis situations. The book provides an overview of different types of emergencies and the unique communication challenges that arise in each situation. It explores the key concepts of emergency communications and the role that ham radio plays in this critical role.

Throughout the book, readers will learn how to prepare themselves and their equipment for emergencies, how to set up their ham radio station for emergency communications, and how to operate during emergencies. The book also covers advanced techniques, such as digital modes, satellite communications, and mesh networking.

The book features real-world case studies of ham radio operators who have played a critical role in emergency communications. These case studies highlight the importance of their work and the lessons that can be learned from their experiences.

The book concludes by looking at emerging trends and technologies in emergency communications, including the role of social media, drones, and other new tools. It explores what the future may hold for ham radio in crisis situations and emphasizes the importance of continued training and practice for ham radio operators.

Overall, this book is an essential guide for anyone interested in emergency communications and the role of ham radio operators in crisis situations. It provides practical advice, tips, and best practices for effective emergency communication and highlights the crucial role that ham radio operators play in ensuring that communication remains available in times of crisis.

1

Introduction

Emergency communications play a crucial role in saving lives and minimizing the impact of disasters. In times of crisis, communication systems can become overloaded, damaged or destroyed, making it difficult or impossible for people to reach their loved ones, call for help, or coordinate response efforts. This is where emergency communications come in.

Effective emergency communication is essential for a timely and coordinated response to emergencies. Emergency communications can help to deliver critical information, coordinate response efforts, and provide vital support to those in need. In many situations, it can be the difference between life and death.

Emergency communications can take many forms, from cell phones and landlines to radios and satellites. While modern communication technologies have made significant progress, they are not always reliable in emergency situations. That's why ham radio operators have long been recognized for their role in providing reliable communication when other means of communication fail or become overloaded.

The importance of emergency communications extends beyond natural disasters and other crises. It includes day-to-day emergencies, such as medical emergencies, car accidents, and fires. Emergency communications are an essential part of public safety, and ham radio operators play a vital role in providing this critical service.

Ham radio plays a vital role in emergency communications. Ham radio operators, also known as amateur radio operators, are licensed by the government to use designated radio frequencies for non-commercial purposes, including emergency communications. In times of crisis, ham radio operators can provide a reliable means of communication when other communication systems have failed or become overloaded.

Role of HAM Radio in Emergency

One of the key benefits of ham radio is its ability to operate independently of traditional communication infrastructure. Ham radio operators can set up their stations in remote or isolated locations and communicate with other operators over long distances, using antennas and other equipment that can be powered by portable generators or batteries. This makes ham radio a valuable tool for emergency responders who need to quickly establish communications in disaster areas or remote locations.

Another benefit of ham radio is its ability to operate on a variety of communication modes. Ham radio operators can communicate using voice, Morse code, digital modes, and more. This versatility makes ham radio a valuable asset in emergency situations where communication needs can vary based on the situation.

Ham radio operators also play an important role in relaying information between emergency responders and those in need. In many cases, ham radio operators can provide critical information about the situation on the ground that can help emergency responders make better decisions and allocate resources more effectively.

Overall, the role of ham radio in emergency communications is essential. It provides a reliable, independent means of communication that can operate when other communication systems have failed, and it helps to ensure that critical information is relayed quickly and efficiently in times of crisis.

Book Outline

This book is a comprehensive guide to emergency communications and the vital role that ham radio operators play in crisis situations. It is divided into ten chapters, each covering different aspects of emergency communications and ham radio operations.

Chapter 1 provides an introduction to the importance of emergency communications and how ham radio fits into this critical role. It also provides an overview of the book's contents.

Chapter 2 explores the different types of emergencies that can occur and the unique communication challenges that arise in each situation. It also introduces the key concepts of emergency communications and the role that ham radio plays.

Chapter 3 provides guidance on how to prepare yourself and your equipment for emergencies. It covers the types of supplies you should have on hand, the importance of training and practice, and the role of emergency plans in ensuring effective communication.

Chapter 4 provides an introduction to ham radio for those who are new to the hobby. It covers information on licensing, equipment, and basic operating procedures, and how these are relevant in emergency communications.

Chapter 5 provides guidance on how to set up your ham radio station for emergency communications. It covers considerations such as power sources, antennas, and communication modes.

Chapter 6 covers the specific operating procedures and best practices for emergency communications. It includes techniques for maximizing signal strength and efficiency, and the use of different communication modes.

Chapter 7 explores the importance of collaboration and teamwork in emergency communications. It covers strategies for coordinating with other operators, first responders, and emergency management officials.

Chapter 8 covers more advanced techniques for emergency communications, such as digital modes, satellite communications, and mesh networking. It explores how these can be used in different types of emergency situations.

Chapter 9 looks at emerging trends and technologies in emergency communications. It covers the role of social media, drones, and other new tools, and what the future may hold for ham radio in crisis situations.

Chapter 10 features real-world examples of ham radio operators who have played a critical role in emergency communications. It highlights the importance of their work and the lessons that can be learned from their experiences.

Overall, this book is a comprehensive guide to emergency communications and the role of ham radio in crisis situations. It provides practical advice, tips, and best practices for effective emergency communication and highlights the crucial role that ham radio operators play in ensuring that communication remains available in times of crisis.

2

Understanding Emergency Communications

Ham Operator

There are many types of emergencies that can occur, each with its unique communication challenges. Here are some of the most common types of emergencies:

1. Natural Disasters - These are events caused by natural phenomena such as earthquakes, hurricanes, tornadoes, floods, landslides, and wildfires. They can cause widespread damage and disruption to infrastructure, including communication systems.
2. Medical Emergencies - These are sudden and unexpected medical events that require urgent care. Examples include heart attacks, strokes, and traumatic injuries.
3. Technological Emergencies - These are events caused by human-made technologies, such as nuclear accidents, chemical spills, and power outages.
4. Civil Unrest - These are events that involve protests, riots, or other forms of civil disturbance. They can create communication challenges as authorities attempt to restore order and protect the public.
5. Cyberattacks - These are attacks on computer systems, networks, and information technology infrastructure. They can cause significant disruption to communication systems and can be difficult to detect and defend against.
6. Terrorist Attacks - These are attacks carried out by individuals or groups with the intention of causing harm, fear, or disruption. They can involve bombings, shootings, and other violent acts, and can create significant communication challenges for emergency responders.
7. Transportation Emergencies - These are events that occur on transportation systems, such as airplane crashes, train derailments, and automobile accidents. They can create communication challenges as responders work to coordinate their efforts and provide assistance to those

affected.

Natural disasters

Natural disasters are events caused by natural phenomena such as weather, geologic, and hydrologic events. They can cause widespread damage and disruption to infrastructure, including communication systems. Here are some common types of natural disasters:

1. Hurricanes - Hurricanes are tropical storms with sustained winds of 74 miles per hour or greater. They can cause extensive damage to coastal areas and inland regions, including flooding and power outages.
2. Tornadoes - Tornadoes are rotating columns of air that can cause significant damage to structures and infrastructure. They can occur in any region of the world but are most common in the central United States.
3. Earthquakes - Earthquakes are the sudden and rapid shaking of the earth's crust caused by the movement of tectonic plates. They can cause significant damage to buildings and infrastructure, as well as landslides and tsunamis.
4. Floods - Floods are caused by heavy rainfall, melting snow, or other events that cause water levels to rise and overflow their banks. They can cause significant damage to buildings and infrastructure, as well as loss of life.
5. Wildfires - Wildfires are uncontrolled fires that occur in areas with combustible vegetation. They can be caused by natural events such as lightning strikes or by human activity. They can cause significant damage to homes and

infrastructure, as well as loss of life.

6. Volcanic Eruptions - Volcanic eruptions occur when magma, ash, and gas are released from a volcano. They can cause significant damage to buildings and infrastructure, as well as ash fall, landslides, and lahars.

In all of these natural disasters, ham radio operators can play a critical role in providing communication links between affected communities and relief agencies. They can set up makeshift radio stations and provide reliable communication when other communication systems have failed or become overloaded.

Man-made emergencies

Man-made emergencies are events that are caused by human activities, either intentionally or unintentionally. They can create significant communication challenges as authorities work to contain the situation and provide assistance to those affected. Here are some common types of man-made emergencies:

1. Industrial Accidents - Industrial accidents can include chemical spills, explosions, and fires that occur in factories, warehouses, or other industrial settings. These incidents can cause significant damage to buildings and infrastructure, as well as environmental damage and health risks.
2. Cyberattacks - Cyberattacks can include viruses, malware, and other forms of cybercrime that target computer systems and networks. These attacks can disrupt communication systems, cause data breaches, and pose significant security risks.

3. Terrorism - Terrorism is the use of violence and intimidation to achieve political or ideological aims. Terrorist attacks can include bombings, shootings, and other violent acts, and can cause significant damage to buildings and infrastructure, as well as loss of life.
4. Transportation Accidents - Transportation accidents can include airplane crashes, train derailments, and automobile accidents. These incidents can cause significant damage to infrastructure, as well as loss of life and injuries.
5. Civil Unrest - Civil unrest can include protests, riots, and other forms of civil disturbance. These incidents can disrupt communication systems and create challenges for authorities as they work to maintain order and protect the public.

In all of these man-made emergencies, ham radio operators can play a critical role in providing communication links between affected communities and relief agencies. They can set up makeshift radio stations and provide reliable communication when other communication systems have failed or become overloaded. Ham radio operators can also help to relay critical information about the situation on the ground and assist with coordinating response efforts.

Communication challenges in emergencies

Emergency situations can create significant communication challenges due to a variety of factors, including infrastructure damage, power outages, and high call volumes. Here are some of the communication challenges that can arise in emergencies:

1. Overloaded Communication Networks - During an emergency, communication networks can quickly become overloaded as people try to contact loved ones and emergency services. This can result in dropped calls, busy signals, and other communication failures.
2. Infrastructure Damage - Natural disasters and other emergencies can cause significant damage to communication infrastructure, including cell towers, power lines, and telephone poles. This can make it difficult or impossible to communicate using traditional communication methods.
3. Power Outages - Power outages can occur during emergencies due to damage to power lines or infrastructure. This can make it difficult or impossible to charge electronic devices and can limit access to communication systems.
4. Language and Cultural Barriers - In emergency situations that involve diverse communities, language and cultural barriers can create challenges for effective communication. It can be difficult to convey critical information to those who do not speak the same language or are not familiar with the culture.
5. Limited Access to Information - In some emergency situations, information may be limited or difficult to obtain, which can create challenges for effective communication. This can be particularly true in situations where authorities are still trying to assess the extent of the damage.
6. Interference from Radio Signals - Radio signals can be interfered with by a variety of factors, including atmospheric conditions, nearby electrical equipment, and interference from other radio signals. This can create communication challenges for ham radio operators and other radio-based communication systems.

In all of these scenarios, ham radio operators can play a critical role in providing reliable communication when other communication systems have failed or become overloaded. By using alternative frequencies and communication modes, ham radio operators can help to ensure that critical information is relayed quickly and efficiently in times of crisis.

Overloaded communication systems

Overloaded communication systems can occur during emergency situations when a large number of people are trying to communicate using the same network. This can cause congestion and result in dropped calls, busy signals, and other communication failures. Here are some of the reasons why communication systems can become overloaded during emergencies:

1. High Call Volumes - During emergencies, people may try to contact loved ones, emergency services, and other important contacts all at once, resulting in high call volumes.
2. Limited Network Capacity - Communication networks can only handle a certain amount of traffic at once. If the volume of calls and messages exceeds the network's capacity, it can become overloaded and fail.
3. Infrastructure Damage - Natural disasters and other emergencies can cause significant damage to communication infrastructure, including cell towers, power lines, and telephone poles. This can limit the capacity of the network and create communication failures.
4. Malicious Activity - In some cases, malicious actors may

try to disrupt communication systems during emergencies, either by launching cyber attacks or by intentionally overwhelming the network with traffic.

To address the problem of overloaded communication systems during emergencies, ham radio operators can provide an alternative means of communication that is not reliant on the traditional communication network. By using ham radio equipment, operators can communicate directly with each other without relying on a central network. This can help to ensure that critical information is relayed quickly and efficiently, even when other communication systems have failed or become overloaded.

Power outages

Power outages are a common occurrence during emergency situations and can create significant communication challenges. Here are some of the reasons why power outages can occur during emergencies:

1. Infrastructure Damage - Natural disasters and other emergencies can cause significant damage to power lines, transformers, and other infrastructure, leading to power outages.
2. Utility Company Precautions - Utility companies may shut off power to certain areas during emergencies as a precautionary measure to prevent electrical fires or other hazards.
3. Fuel Shortages - Fuel shortages can occur during emergencies, which can make it difficult to generate electricity

and maintain power systems.

Power outages can create communication challenges in several ways, including:

1. Inability to Charge Devices - During power outages, it can be difficult or impossible to charge electronic devices, which can limit access to communication systems.
2. Loss of Internet Access - Many communication systems, including email and social media, require an internet connection. If the power outage also affects internet infrastructure, it can limit access to these communication systems.
3. Loss of Traditional Communication Methods - During power outages, landlines and cellular networks may become unavailable, which can limit access to traditional communication methods.

To address the problem of power outages during emergencies, ham radio operators can use alternative power sources, such as batteries or generators, to power their equipment. This can help to ensure that communication systems remain available even in the event of a power outage. Ham radio operators can also use low-power communication modes, such as Morse code or digital modes, to conserve battery power and maximize the efficiency of their communication systems.

Language and Cultural Barriers

During emergency situations that involve diverse communities, language and cultural barriers can create challenges for effective communication. Here are some of the ways that language and cultural barriers can affect emergency communication:

1. Limited Language Proficiency - In emergency situations, people who do not speak the local language may have difficulty communicating with emergency responders and accessing critical information.
2. Different Communication Styles - Different cultures may have different communication styles, which can create challenges for effective communication during emergencies.
3. Misunderstandings - Language and cultural barriers can create misunderstandings that can lead to confusion and errors in communication.
4. Access to Information - In some emergency situations, information may only be available in certain languages, which can limit access to critical information for people who do not speak those languages.

To address these challenges, emergency responders and communication operators can take a number of steps, including:

1. Providing Language Support - Providing language support through interpreters, translation services, or multilingual communication equipment can help to ensure that all members of a community have access to critical information.

2. Cultural Competency Training - Cultural competency training can help emergency responders and communication operators to better understand and navigate cultural differences in communication styles and expectations.
3. Accessible Communication - Ensuring that critical information is accessible in multiple languages and communication modes can help to ensure that all members of a community can access important information.
4. Collaboration with Community Organizations - Collaborating with community organizations that serve diverse populations can help emergency responders and communication operators to better understand and respond to the needs of these populations during emergencies.

Limited Access to Information

In some emergency situations, information may be limited or difficult to obtain, which can create challenges for effective communication. Here are some of the ways that limited access to information can affect emergency communication:

1. Uncertainty - When information is limited, people may not know what is happening or what actions to take, which can create fear and uncertainty.
2. Delayed Response - Without accurate and timely information, emergency responders may not be able to respond as quickly or effectively as possible.
3. Misinformation - Limited access to information can lead to misinformation and rumors, which can create confusion and panic.
4. Lack of Preparedness - When people are not aware of

potential risks or emergency procedures, they may not be prepared to respond effectively in emergency situations.

To address these challenges, emergency responders and communication operators can take a number of steps, including:

1. Information Sharing - Sharing information between emergency responders, government agencies, and the public can help to ensure that accurate and timely information is available to everyone.
2. Risk Communication - Communicating the risks associated with emergency situations and providing guidance on appropriate responses can help to reduce fear and uncertainty.
3. Multi-Channel Communication - Using multiple communication channels, such as social media, text messaging, and traditional media, can help to ensure that critical information reaches as many people as possible.
4. Emergency Preparedness Education - Providing education and training on emergency preparedness can help to ensure that people are better prepared to respond effectively in emergency situations, even when information is limited.

Ham radio operators can also play a critical role in providing reliable communication links during emergencies when information is limited. By using alternative frequencies and communication modes, ham radio operators can help to ensure that critical information is relayed quickly and efficiently, even when other communication systems have failed or become overloaded.

Interference from Radio Signals

Radio signals can be interfered with by a variety of factors, including atmospheric conditions, nearby electrical equipment, and interference from other radio signals. This can create communication challenges for ham radio operators and other radio-based communication systems. Here are some of the ways that radio signal interference can affect emergency communication:

1. Weak Signal Strength - Interference can weaken radio signals, making it difficult to transmit and receive clear signals.
2. Noise and Distortion - Interference can cause noise and distortion in radio signals, making it difficult to understand the information being transmitted.
3. Signal Blocking - Interference can block radio signals entirely, preventing communication from taking place.
4. Inconsistent Signal Quality - Interference can cause signal quality to vary, making it difficult to maintain a reliable communication link.

To address these challenges, ham radio operators can take a number of steps, including:

1. Use Alternative Frequencies - When a frequency is experiencing interference, ham radio operators can switch to alternative frequencies to avoid interference.
2. Use Filtering Equipment - Filtering equipment, such as noise filters and signal boosters, can help to reduce interference and improve signal quality.
3. Use Directional Antennas - Directional antennas can help

to focus radio signals in a specific direction, reducing the impact of interference from other sources.

4. Practice Good Operating Techniques - Good operating techniques, such as using proper microphone placement and speaking clearly and slowly, can help to improve signal quality and reduce interference.

By using these techniques, ham radio operators can help to ensure that critical information is relayed quickly and efficiently during emergency situations, even when radio signal interference is present.

Key concepts of emergency communications

Emergency communications is a specialized field that involves the use of communication systems to support emergency response activities. Here are some of the key concepts of emergency communications:

1. Interoperability - Interoperability refers to the ability of different communication systems to work together seamlessly. In emergency situations, it is critical that different agencies and organizations can communicate with each other effectively, even if they are using different communication systems.
2. Redundancy - Redundancy refers to the use of multiple communication systems to ensure that communication links remain available even in the event of a failure or disruption in one system. Redundancy can include the use of backup power supplies, alternative communication modes, and alternative frequencies.

3. Priority Communications - Priority communications refer to the use of communication systems to support emergency response activities, such as directing emergency responders to the scene of an incident, coordinating evacuation efforts, and providing critical updates on the situation.
4. Standard Operating Procedures - Standard Operating Procedures (SOPs) are established procedures and protocols that are designed to ensure that communication systems are used effectively and efficiently during emergency situations. SOPs can include procedures for contacting emergency services, establishing communication links, and managing communication traffic.
5. Training and Practice - Training and practice are critical to effective emergency communications. Operators must be trained on the proper use of communication equipment, as well as the procedures and protocols that are used during emergency situations. Regular practice and drills can help to ensure that operators are prepared to respond effectively in a crisis.

In all of these areas, ham radio operators can play a critical role in providing reliable communication links during emergencies. By using alternative frequencies and communication modes, ham radio operators can help to ensure that critical information is relayed quickly and efficiently, even when other communication systems have failed or become overloaded.

Interoperability

Interoperability is the ability of different communication systems to work together seamlessly. In emergency situations, it is critical that different agencies and organizations can communicate with each other effectively, even if they are using different communication systems. Here are some of the ways that interoperability can affect emergency communication:

1. Collaborative Communication - Interoperability allows different agencies and organizations to communicate collaboratively during emergency situations. This can help to ensure that everyone involved in the response effort has access to critical information.
2. Efficient Resource Allocation - Interoperability can help emergency responders to allocate resources more efficiently by allowing them to communicate more effectively with other agencies and organizations.
3. Reduced Response Time - Interoperability can help to reduce response time by allowing emergency responders to communicate more quickly and effectively with each other.
4. Improved Situational Awareness - Interoperability can help to improve situational awareness by providing emergency responders with a more comprehensive view of the situation.

To address interoperability challenges, emergency responders and communication operators can take a number of steps, including:

1. Establish Common Communication Protocols - Establishing common communication protocols can help to ensure that different communication systems can work together seamlessly.
2. Conduct Interoperability Testing - Conducting interoperability testing can help to identify any issues with communication systems before an emergency situation occurs.
3. Use Multi-Agency Communication Equipment - Using multi-agency communication equipment, such as interoperable radio systems, can help to ensure that emergency responders can communicate effectively with each other.
4. Collaborate with Other Agencies - Collaborating with other agencies and organizations during emergency situations can help to ensure that everyone involved in the response effort is working towards a common goal.

By addressing interoperability challenges, emergency responders and communication operators can help to ensure that critical information is relayed quickly and efficiently during emergency situations. This can help to improve the overall response effort and save lives.

Redundancy

Redundancy is the use of multiple communication systems to ensure that communication links remain available even in the event of a failure or disruption in one system. In emergency situations, redundancy is critical to ensuring that critical information is relayed quickly and efficiently. Here are some of the ways that redundancy can affect emergency communication:

1. Backup Communication Options - Having backup communication options, such as satellite phones or ham radio equipment, can help to ensure that communication links remain available even if traditional communication systems have failed.
2. Alternative Power Sources - Having alternative power sources, such as generators or battery packs, can help to ensure that communication equipment remains operational even if the main power supply has failed.
3. Multi-Channel Communication - Using multiple communication channels, such as landlines, cellular networks, and ham radio equipment, can help to ensure that critical information is relayed even if one channel has failed.
4. Cross-Training - Cross-training personnel on different communication systems can help to ensure that there are backup operators available to maintain communication links if one operator is unavailable.

To address redundancy challenges, emergency responders and communication operators can take a number of steps, including:

1. Conduct Risk Assessments - Conducting risk assessments can help to identify potential failure points in communication systems and develop backup plans.
2. Test Backup Systems - Regularly testing backup systems can help to ensure that they are operational and effective in emergency situations.
3. Collaborate with Other Agencies - Collaborating with other agencies and organizations can help to ensure that redundancy plans are coordinated and effective.

4. Maintain Communication Equipment - Maintaining communication equipment and ensuring that backup equipment is available and operational can help to ensure that redundancy plans are effective.

By addressing redundancy challenges, emergency responders and communication operators can help to ensure that critical information is relayed quickly and efficiently during emergency situations. This can help to improve the overall response effort and save lives.

Priority Communications

Priority communications refer to the use of communication systems to support emergency response activities, such as directing emergency responders to the scene of an incident, coordinating evacuation efforts, and providing critical updates on the situation. Here are some of the ways that priority communications can affect emergency communication:

1. Faster Response Time - Priority communications can help emergency responders to respond more quickly to emergencies, which can save lives and prevent further damage.
2. Coordination of Response Efforts - Priority communications can help to coordinate response efforts among different agencies and organizations involved in emergency response activities.
3. Situational Awareness - Priority communications can help to provide emergency responders with a more comprehensive view of the situation, which can help to improve

situational awareness and decision-making.

4. Clear and Accurate Communication - Priority communications can help to ensure that critical information is communicated clearly and accurately, which can help to avoid misunderstandings and errors in response efforts.

To address priority communications challenges, emergency responders and communication operators can take a number of steps, including:

1. Establish Priority Communication Protocols - Establishing priority communication protocols can help to ensure that critical information is communicated quickly and effectively.
2. Use Priority Communication Channels - Using priority communication channels, such as dedicated radio frequencies or emergency notification systems, can help to ensure that critical information is relayed quickly and efficiently.
3. Train Personnel on Priority Communications - Providing training on priority communications to emergency responders and communication operators can help to ensure that they are prepared to respond effectively in emergency situations.
4. Use Clear and Concise Language - Using clear and concise language in priority communications can help to ensure that critical information is communicated accurately and quickly.

By addressing priority communications challenges, emergency responders and communication operators can help to ensure that critical information is relayed quickly and efficiently during emergency situations. This can help to improve the overall response effort and save lives.

Standard Operating Procedures

Standard operating procedures (SOPs) are guidelines for how communication systems should be used during emergency situations. SOPs can help to ensure that communication systems are used effectively and efficiently, and that all personnel involved in emergency response efforts are working towards a common goal. Here are some of the ways that SOPs can affect emergency communication:

1. Consistency - SOPs help to ensure that communication systems are used consistently across different agencies and organizations involved in emergency response efforts.
2. Efficiency - SOPs can help to streamline communication processes, which can reduce response time and improve the overall efficiency of emergency response efforts.
3. Safety - SOPs can help to ensure that communication systems are used safely and effectively, which can help to prevent injuries and damage during emergency situations.
4. Coordination - SOPs can help to coordinate communication efforts among different agencies and organizations involved in emergency response efforts, which can help to improve the overall response effort.

To address SOP challenges, emergency responders and communication operators can take a number of steps, including:

1. Develop SOPs - Developing SOPs specific to the communi-

cation systems used during emergency response efforts can help to ensure that communication systems are used effectively and efficiently.

2. Train Personnel on SOPs - Providing training on SOPs to emergency responders and communication operators can help to ensure that they are familiar with the guidelines and prepared to use them during emergency situations.
3. Review and Update SOPs - Regularly reviewing and updating SOPs can help to ensure that they remain relevant and effective in emergency situations.
4. Collaborate with Other Agencies - Collaborating with other agencies and organizations involved in emergency response efforts can help to ensure that SOPs are coordinated and effective across all involved parties.

By addressing SOP challenges, emergency responders and communication operators can help to ensure that communication systems are used effectively and efficiently during emergency situations. This can help to improve the overall response effort and save lives.

Training and Practice

Training and practice are critical to effective emergency communication. In order to respond effectively to emergency situations, emergency responders and communication operators must be familiar with the communication systems and equipment that they will use during emergency response efforts. Here are some of the ways that training and practice can affect emergency communication:

1. Familiarity - Training and practice can help emergency responders and communication operators become familiar with communication systems and equipment, which can help to reduce response time and improve the overall efficiency of emergency response efforts.
2. Confidence - Training and practice can help to build confidence among emergency responders and communication operators, which can help to improve their ability to respond effectively in emergency situations.
3. Coordination - Training and practice can help to improve coordination among different agencies and organizations involved in emergency response efforts, which can help to improve the overall response effort.
4. Safety - Training and practice can help to ensure that communication systems and equipment are used safely and effectively, which can help to prevent injuries and damage during emergency situations.

To address training and practice challenges, emergency responders and communication operators can take a number of steps, including:

1. Provide Training - Providing training on communication systems and equipment can help to ensure that emergency responders and communication operators are familiar with the systems and prepared to use them effectively in emergency situations.
2. Conduct Drills and Exercises - Conducting drills and exercises can help to simulate emergency situations and provide opportunities for emergency responders and communication operators to practice their skills.

3. Participate in Training Programs - Participating in training programs offered by government agencies or other organizations can help to ensure that emergency responders and communication operators are familiar with the latest communication technologies and best practices.
4. Conduct After-Action Reviews - Conducting after-action reviews following emergency situations can help to identify areas for improvement in communication systems and equipment and inform future training efforts.

By addressing training and practice challenges, emergency responders and communication operators can help to ensure that communication systems are used effectively and efficiently during emergency situations. This can help to improve the overall response effort and save lives.

3

Preparing for Emergencies

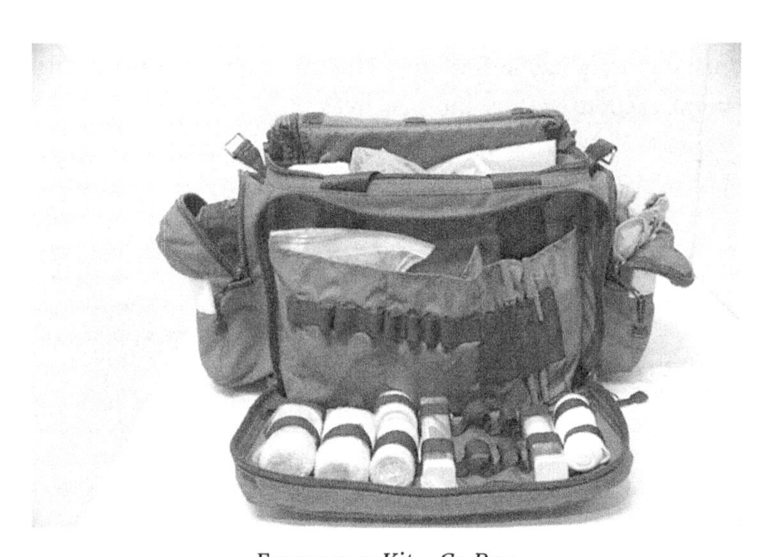

Emergency Kit - Go Bag

Personal preparation is critical to effective emergency communication. In order to respond effectively to emergency situations, emergency responders and communication operators must be prepared to operate in a variety of challenging environments and conditions. Here are some of the ways that personal preparation can affect emergency communication:

1. Physical Fitness - Emergency responders and communication operators must be physically fit in order to respond effectively to emergency situations. Maintaining good physical health can help to improve endurance, reduce fatigue, and improve overall performance during emergency situations.
2. Mental Preparation - Emergency responders and communication operators must be mentally prepared to respond to emergency situations. This includes being able to remain calm and focused in high-stress situations and being able to make quick decisions under pressure.
3. Equipment Preparation - Emergency responders and communication operators must ensure that their equipment is prepared and ready to use during emergency situations. This includes ensuring that equipment is properly maintained, charged, and ready to use.
4. Communication Preparation - Emergency responders and communication operators must be prepared to communicate effectively during emergency situations. This includes being familiar with communication protocols, practicing good communication skills, and being able to adapt to changing communication environments.

To address personal preparation challenges, emergency responders and communication operators can take a number of steps, including:

1. Maintain Physical Fitness - Maintaining good physical health through regular exercise and healthy eating can help to ensure that emergency responders and communication operators are physically prepared to respond to emergency situations.
2. Participate in Mental Health Training - Participating in mental health training, such as stress management or resiliency training, can help to ensure that emergency responders and communication operators are mentally prepared to respond to emergency situations.
3. Conduct Equipment Checks - Regularly checking and maintaining equipment can help to ensure that it is prepared and ready to use during emergency situations.
4. Participate in Communication Training - Participating in communication training, such as communication protocol training or public speaking training, can help to ensure that emergency responders and communication operators are prepared to communicate effectively during emergency situations.

By addressing personal preparation challenges, emergency responders and communication operators can help to ensure that they are prepared to respond effectively to emergency situations. This can help to improve the overall response effort and save lives.

Emergency Kits

Emergency kits are essential to have in emergency situations. Emergency kits should contain items that can help to sustain life and provide essential supplies during and after emergency situations. Here are some of the items that should be included in an emergency kit:

1. Water - Emergency kits should contain at least one gallon of water per person per day for at least three days.
2. Non-perishable food - Emergency kits should contain at least a three-day supply of non-perishable food, such as canned goods, protein bars, and other non-perishable food items.
3. First Aid Kit - An emergency kit should contain a first aid kit with basic medical supplies, including bandages, antiseptic, pain relief medication, and other basic medical supplies.
4. Emergency Blanket - An emergency blanket can be used to keep a person warm in cold environments or to protect them from the elements.
5. Flashlight - An emergency kit should contain a flashlight with extra batteries or a hand-cranked flashlight.
6. Radio - An emergency kit should contain a battery-powered or hand-cranked radio to receive updates and emergency information.
7. Multi-tool - An emergency kit should contain a multi-tool, such as a Swiss Army Knife, that includes basic tools like a knife, scissors, and pliers.
8. Whistle - An emergency kit should contain a whistle to signal for help or to alert others to your location during

emergency situations.

9. Personal Hygiene Items - An emergency kit should contain personal hygiene items, such as toothbrushes, toothpaste, and hand sanitizer.
10. Copies of important documents - An emergency kit should contain copies of important documents, such as identification documents, insurance policies, and other important paperwork.

By including these items in an emergency kit, emergency responders and communication operators can help to provide essential supplies and sustain life during emergency situations. This can help to improve the overall response effort and save lives.

First Aid Kit

A first aid kit is an essential item to have in emergency situations. Injuries and medical emergencies can occur during emergency situations, and having a first aid kit on hand can help to provide immediate care until medical professionals arrive. Here are some of the items that should be included in a basic first aid kit:

1. Bandages - A variety of bandages, including adhesive bandages, sterile gauze, and self-adhesive wrap, should be included in a first aid kit to cover and protect cuts, scrapes, and other injuries.
2. Antiseptic - Antiseptic wipes or ointment can be used to clean wounds and prevent infection.
3. Pain Relief - Pain relief medication, such as ac-

etaminophen or ibuprofen, can help to relieve pain and reduce fever.

4. Tweezers - Tweezers can be used to remove splinters or foreign objects from wounds.
5. Scissors - Scissors can be used to cut bandages, clothing, or other materials.
6. Gloves - Disposable gloves can help to prevent the spread of infection during first aid treatment.
7. Emergency Blanket - An emergency blanket can be used to keep a person warm in cold environments or to protect them from the elements.
8. CPR Mask - A CPR mask can be used to provide mouth-to-mouth resuscitation during an emergency situation.
9. Flashlight - A flashlight can be used to provide illumination during emergency situations where there is no power.
10. Whistle - A whistle can be used to signal for help or to alert others to your location during emergency situations.

By including these items in a first aid kit, emergency responders and communication operators can help to provide immediate care during emergency situations, which can help to improve the overall response effort and save lives.

First Aid Training

First aid training is essential for emergency responders and communication operators. Injuries and medical emergencies can occur during emergency situations, and having the knowledge and skills to provide immediate care can help to save lives. Here are some of the ways that first aid training can affect emergency communication:

1. Improved Response Time - By having the knowledge and skills to provide immediate care, emergency responders and communication operators can improve their response time during emergency situations.
2. Increased Confidence - First aid training can help to build confidence among emergency responders and communication operators, which can help to improve their ability to respond effectively in emergency situations.
3. Better Communication - First aid training can help emergency responders and communication operators communicate more effectively with each other, medical professionals, and others involved in emergency response efforts.
4. Enhanced Safety - First aid training can help to ensure that emergency responders and communication operators are able to provide care safely and effectively, which can help to prevent injuries and damage during emergency situations.

To address first aid training challenges, emergency responders and communication operators can take a number of steps, including:

1. Attend First Aid Training Courses - Attending first aid training courses can help to provide emergency responders and communication operators with the knowledge and skills necessary to provide immediate care during emergency situations.
2. Practice First Aid Techniques - Practicing first aid techniques can help to improve emergency responders' and communication operators' ability to respond effectively

in emergency situations.

3. Participate in Refresher Courses - Participating in refresher courses can help to ensure that emergency responders and communication operators remain up-to-date on the latest first aid techniques and best practices.
4. Encourage Ongoing Learning - Encouraging ongoing learning can help to ensure that emergency responders and communication operators remain informed about new developments in first aid and emergency response.

By addressing first aid training challenges, emergency responders and communication operators can help to ensure that they are able to provide immediate care during emergency situations. This can help to improve the overall response effort and save lives.

Equipment preparation

Equipment preparation is crucial for emergency communication operators. Communication equipment must be in good working condition and ready to use during emergency situations. Here are some of the ways that equipment preparation can affect emergency communication:

1. Quick Response - Properly prepared equipment can help emergency responders and communication operators to respond quickly to emergency situations.
2. Improved Communication - Prepared equipment can help to ensure that communication systems and equipment are in good working condition and ready to use, which can help to improve communication during emergency

situations.

3. Safety - Properly prepared equipment can help to prevent equipment failure or damage, which can improve safety during emergency situations.
4. Increased Efficiency - Prepared equipment can help to increase the efficiency of communication during emergency situations, allowing for more effective communication among emergency responders.

To address equipment preparation challenges, emergency responders and communication operators can take a number of steps, including:

1. Regular Maintenance - Regular maintenance of communication equipment can help to ensure that it is in good working condition and ready to use during emergency situations.
2. Testing - Testing communication equipment regularly can help to identify any issues and ensure that it is working properly.
3. Backup Equipment - Having backup equipment, such as backup radios or power sources, can help to ensure that communication equipment is available during emergency situations.
4. Equipment Inventory - Keeping an inventory of communication equipment can help to ensure that all equipment is accounted for and in good working condition.

By addressing equipment preparation challenges, emergency responders and communication operators can help to ensure that communication equipment is ready to use during emergency situations. This can help to improve the overall response effort and save lives.

Backup power sources

Backup power sources are essential for emergency communication operators. Power outages can occur during emergency situations, which can cause communication equipment to fail. Backup power sources can help to ensure that communication equipment remains operational during power outages. Here are some of the backup power sources that can be used in emergency situations:

1. Generators - Generators are a common backup power source and can provide power for extended periods of time. Portable generators can be used in the field, while stationary generators can be used at fixed locations.
2. Battery Banks - Battery banks can be used to store energy and provide power during power outages. They can be charged with solar panels or other alternative power sources.
3. Uninterruptible Power Supplies (UPS) - UPS systems can provide backup power for a limited period of time, typically between 15-30 minutes. They are ideal for providing temporary power while backup power sources are being activated.
4. Fuel Cells - Fuel cells convert chemical energy into electrical energy and can provide backup power for extended periods of time. They are an efficient and environmentally friendly alternative to traditional generators.

To address backup power source challenges, emergency responders and communication operators can take a number of steps, including:

1. Determine Power Needs - Determining power needs can help to identify the type and amount of backup power sources needed for emergency communication equipment.
2. Regular Maintenance - Regular maintenance of backup power sources can help to ensure that they are in good working condition and ready to use during emergency situations.
3. Test Backup Power Sources - Testing backup power sources regularly can help to ensure that they are working properly and ready to use during emergency situations.
4. Backup Power Source Inventory - Keeping an inventory of backup power sources can help to ensure that all equipment is accounted for and in good working condition.

By addressing backup power source challenges, emergency responders and communication operators can help to ensure that communication equipment remains operational during power outages. This can help to improve the overall response effort and save lives.

Portable antennas

Portable antennas are important for emergency communication operators who need to set up communication systems quickly and efficiently in the field. Portable antennas can be easily transported to different locations and can be set up in a variety of configurations depending on the communication needs. Here are some of the types of portable antennas that can be used in emergency situations:

1. Vertical Antennas - Vertical antennas are easy to set up and can be used for both short-range and long-range communication. They are suitable for a wide range of frequencies and can be used in a variety of environments.
2. Dipole Antennas - Dipole antennas are simple and easy to set up, and can be used for both short-range and long-range communication. They are ideal for emergency situations where communication needs to be established quickly.
3. Yagi Antennas - Yagi antennas are directional antennas that can be used for long-range communication. They are ideal for emergency situations where communication needs to be established over long distances.
4. Magnetic Loop Antennas - Magnetic loop antennas are small and portable and can be used for communication in areas where space is limited or where the environment is noisy.

To address portable antenna challenges, emergency responders and communication operators can take a number of steps, including:

1. Determine Antenna Needs - Determining antenna needs can help to identify the type and configuration of portable antennas needed for emergency communication equipment.
2. Practice Setup - Practicing antenna setup can help to ensure that emergency responders and communication

operators can set up portable antennas quickly and efficiently during emergency situations.

3. Test Antennas - Testing antennas regularly can help to ensure that they are working properly and ready to use during emergency situations.
4. Antenna Inventory - Keeping an inventory of portable antennas can help to ensure that all equipment is accounted for and in good working condition.

By addressing portable antenna challenges, emergency responders and communication operators can help to ensure that communication equipment is set up quickly and efficiently during emergency situations. This can help to improve the overall response effort and save lives.

Emergency Plans

Emergency plans are essential for emergency communication operators. Emergency plans help to ensure that emergency responders and communication operators are prepared for a variety of emergency situations and have a plan of action in place to respond effectively. Here are some of the components of an effective emergency plan:

1. Risk Assessment - A risk assessment should be conducted to identify potential emergency situations that could occur and their likelihood.
2. Communication Protocols - Communication protocols should be established to ensure that emergency responders and communication operators can communicate effectively during emergency situations.

3. Emergency Notification Procedures - Emergency notification procedures should be established to ensure that all necessary personnel are notified in the event of an emergency.
4. Contingency Planning - Contingency planning should be conducted to ensure that backup systems and procedures are in place in the event of equipment failure or other emergency situations.
5. Resource Management - Resource management procedures should be established to ensure that all necessary resources, including personnel and equipment, are available during emergency situations.

To address emergency plan challenges, emergency responders and communication operators can take a number of steps, including:

1. Regular Plan Review - Emergency plans should be reviewed regularly to ensure that they remain current and effective.
2. Practice Scenarios - Practicing emergency scenarios can help to ensure that emergency responders and communication operators are prepared to respond effectively during emergency situations.
3. Refine Procedures - Refining emergency procedures based on lessons learned from previous emergency situations can help to improve the overall response effort.
4. Communication - Communication among emergency responders and communication operators can help to ensure that all necessary personnel are informed of the emergency plan and their roles during emergency situations.

By addressing emergency plan challenges, emergency responders and communication operators can help to ensure that they are prepared to respond effectively during emergency situations. This can help to improve the overall response effort and save lives.

Risk Assessment

Risk assessment is an essential component of emergency planning for emergency communication operators. A risk assessment helps to identify potential emergency situations that could occur and their likelihood, which can help to inform emergency planning efforts. Here are some of the steps that emergency communication operators can take to conduct a risk assessment:

1. Identify Potential Emergency Situations - The first step in conducting a risk assessment is to identify potential emergency situations that could occur. This could include natural disasters, man-made disasters, and other emergencies that could impact communication systems.
2. Determine Likelihood - The likelihood of each potential emergency situation should be determined. This can be based on historical data, current conditions, and other factors.
3. Assess Impact - The potential impact of each emergency situation should be assessed. This could include the impact on communication systems, emergency response efforts, and the community as a whole.
4. Identify Vulnerabilities - Vulnerabilities in communication systems and emergency response efforts should be

identified. This can include equipment vulnerabilities, personnel vulnerabilities, and other factors that could impact emergency response efforts.

5. Develop Mitigation Strategies - Mitigation strategies should be developed to address the vulnerabilities identified in the risk assessment. This could include backup systems, emergency procedures, and other strategies to mitigate the impact of potential emergency situations.

To address risk assessment challenges, emergency communication operators can take a number of steps, including:

1. Regular Review - Risk assessments should be reviewed regularly to ensure that they remain current and reflect changing conditions.
2. Collaboration - Collaboration among emergency communication operators and other emergency responders can help to ensure that all potential emergency situations are considered in the risk assessment.
3. Refine Strategies - Risk assessments should inform emergency planning efforts and help to refine mitigation strategies based on lessons learned from previous emergency situations.

By conducting a risk assessment and addressing vulnerabilities, emergency communication operators can help to ensure that they are prepared to respond effectively during emergency situations. This can help to improve the overall response effort and save lives.

Communication Protocols

Communication protocols are an essential component of emergency planning for emergency communication operators. Communication protocols help to ensure that emergency responders and communication operators can communicate effectively during emergency situations. Here are some of the steps that emergency communication operators can take to establish effective communication protocols:

1. Identify Communication Needs - The first step in establishing effective communication protocols is to identify communication needs. This can include the types of communication that will be needed, the frequency of communication, and the channels that will be used.
2. Establish Communication Hierarchy - A communication hierarchy should be established to ensure that all necessary personnel are informed of the emergency situation and their roles in the response effort.
3. Define Communication Procedures - Communication procedures should be defined to ensure that emergency responders and communication operators are communicating effectively and efficiently during emergency situations.
4. Use Standard Terminology - Standard terminology should be used to ensure that all emergency responders and communication operators understand the information being communicated.
5. Conduct Training and Exercises - Training and exercises should be conducted to ensure that all personnel understand the communication protocols and can communicate

effectively during emergency situations.

To address communication protocol challenges, emergency communication operators can take a number of steps, including:

1. Regular Review - Communication protocols should be reviewed regularly to ensure that they remain current and effective.
2. Collaborate with Emergency Responders - Collaboration with emergency responders can help to ensure that all communication needs are identified and that the communication hierarchy is established.
3. Practice Scenarios - Practicing emergency scenarios can help to ensure that emergency responders and communication operators are prepared to communicate effectively during emergency situations.

By establishing effective communication protocols and addressing communication protocol challenges, emergency communication operators can help to ensure that they are prepared to communicate effectively during emergency situations. This can help to improve the overall response effort and save lives.

Emergency Notification Procedures

Emergency notification procedures are an essential component of emergency planning for emergency communication operators. Emergency notification procedures help to ensure that all necessary personnel are notified in the event of an emergency. Here are some of the steps that emergency communication operators can take to establish effective emergency notification procedures:

1. Identify Key Personnel - The first step in establishing effective emergency notification procedures is to identify key personnel who need to be notified in the event of an emergency.
2. Establish Notification Methods - Notification methods should be established to ensure that all necessary personnel can be notified quickly and effectively. This can include phone calls, text messages, emails, and other methods of communication.
3. Define Notification Procedures - Notification procedures should be defined to ensure that emergency responders and communication operators can quickly and efficiently notify all necessary personnel in the event of an emergency.
4. Test Notification Procedures - Notification procedures should be tested regularly to ensure that they are working properly and that all necessary personnel can be notified quickly and effectively.
5. Use Redundant Notification Methods - Redundant notification methods should be used to ensure that all necessary personnel are notified in the event of a failure of the primary notification method.

To address emergency notification procedure challenges, emergency communication operators can take a number of steps, including:

1. Regular Review - Emergency notification procedures

should be reviewed regularly to ensure that they remain current and effective.

2. Collaborate with Emergency Responders - Collaboration with emergency responders can help to ensure that all key personnel are identified and that notification procedures are established.
3. Conduct Training and Exercises - Training and exercises should be conducted to ensure that all personnel understand the notification procedures and can respond effectively during emergency situations.

By establishing effective emergency notification procedures and addressing emergency notification procedure challenges, emergency communication operators can help to ensure that all necessary personnel are notified quickly and effectively in the event of an emergency. This can help to improve the overall response effort and save lives.

Contingency Planning

Contingency planning is an essential component of emergency planning for emergency communication operators. Contingency planning helps to ensure that backup systems and procedures are in place in the event of equipment failure or other emergency situations. Here are some of the steps that emergency communication operators can take to establish effective contingency plans:

1. Identify Critical Systems - The first step in establishing effective contingency plans is to identify critical systems that must be maintained during emergency situations.

This can include communication systems, power systems, and other critical infrastructure.

2. Establish Backup Systems - Backup systems should be established to ensure that critical systems can be maintained in the event of equipment failure or other emergency situations. This can include backup power sources, backup communication systems, and other redundant systems.
3. Develop Standard Operating Procedures - Standard operating procedures should be developed to ensure that emergency responders and communication operators can quickly and efficiently switch to backup systems in the event of an emergency.
4. Test Contingency Plans - Contingency plans should be tested regularly to ensure that they are working properly and that all necessary personnel understand their roles in implementing the plans.
5. Develop Recovery Procedures - Recovery procedures should be developed to ensure that critical systems can be restored as quickly as possible after an emergency situation.

To address contingency planning challenges, emergency communication operators can take a number of steps, including:

1. Regular Review - Contingency plans should be reviewed regularly to ensure that they remain current and effective.
2. Collaboration with Emergency Responders - Collaboration with emergency responders can help to ensure that critical systems are identified and that backup systems are established to maintain these systems during emergency situations.

3. Conduct Training and Exercises - Training and exercises should be conducted to ensure that all personnel understand the contingency plans and can respond effectively during emergency situations.

By establishing effective contingency plans and addressing contingency planning challenges, emergency communication operators can help to ensure that critical systems are maintained and that emergency response efforts can be effectively coordinated during emergency situations. This can help to improve the overall response effort and save lives.

Resource Management

Resource management is an essential component of emergency planning for emergency communication operators. Resource management helps to ensure that necessary resources, such as equipment, supplies, and personnel, are available and properly allocated during emergency situations. Here are some of the steps that emergency communication operators can take to establish effective resource management:

1. Identify Necessary Resources - The first step in establishing effective resource management is to identify the necessary resources that will be needed during emergency situations. This can include communication equipment, power sources, supplies, and personnel.
2. Establish Resource Inventory - A resource inventory should be established to track the availability of necessary resources. This can include the quantity and location of resources, as well as the condition and maintenance status

of equipment.

3. Develop Allocation Procedures - Procedures should be developed to ensure that necessary resources are allocated to emergency responders and communication operators as needed during emergency situations.
4. Establish Resource Tracking - A system for tracking the allocation and use of resources should be established to ensure that resources are properly managed and accounted for.
5. Conduct Regular Maintenance - Regular maintenance and inspections of equipment and supplies should be conducted to ensure that resources are in good condition and available for use during emergency situations.

To address resource management challenges, emergency communication operators can take a number of steps, including:

1. Regular Review - Resource management plans and procedures should be reviewed regularly to ensure that they remain current and effective.
2. Collaboration with Emergency Responders - Collaboration with emergency responders can help to ensure that necessary resources are identified and that procedures for allocating resources are established.
3. Conduct Training and Exercises - Training and exercises should be conducted to ensure that all personnel understand the resource management procedures and can respond effectively during emergency situations.

By establishing effective resource management and addressing resource management challenges, emergency communication operators can help to ensure that necessary resources are available and properly allocated during emergency situations. This can help to improve the overall response effort and save lives.

Practice drills

Practice drills are an essential component of emergency planning for emergency communication operators. Practice drills help to ensure that all personnel understand their roles and responsibilities during emergency situations and can respond effectively. Here are some of the steps that emergency communication operators can take to establish effective practice drills:

1. Establish Practice Drill Schedule - A schedule for conducting practice drills should be established to ensure that all personnel have an opportunity to participate in drills.
2. Define Practice Drill Scenarios - Scenarios for practice drills should be defined to ensure that all types of emergency situations are covered.
3. Conduct Practice Drills - Practice drills should be conducted regularly to ensure that all personnel understand their roles and responsibilities during emergency situations and can respond effectively.
4. Evaluate Practice Drills - Practice drills should be evaluated to identify areas for improvement and to ensure that all personnel are adequately prepared for emergency situations.
5. Adjust Emergency Plans - Emergency plans should be adjusted based on the results of practice drills to ensure

that they are effective and up-to-date.

To address practice drill challenges, emergency communication operators can take a number of steps, including:

1. Regular Review - Practice drills should be reviewed regularly to ensure that they remain current and effective.
2. Collaboration with Emergency Responders - Collaboration with emergency responders can help to ensure that practice drills are tailored to the specific needs of the community.
3. Conduct Training - Training should be conducted to ensure that all personnel understand the purpose and importance of practice drills and are prepared to participate effectively.

By establishing effective practice drills and addressing practice drill challenges, emergency communication operators can help to ensure that all personnel are prepared to respond effectively during emergency situations. This can help to improve the overall response effort and save lives.

4

Getting Started with Ham Radio

CW ONLY STATION
5 BAND DXCC CW

N7RD

Confirming QSO with	Date			UTC	MHz	RST	CW	QSL
	Month	Day	Year					
								PSE TNX

Grid Sq. DM33uq
CWops #1551
SKCC #5507T
Fists #2714

Ron D. Smith
12408 W. Candlelight Dr
Sun City West, AZ.
85375-3342 USA

QSL by N7RD

Typical Amateur Radio License

Licensing requirements are an important aspect of ham radio operation. The Federal Communications Commission (FCC) regulates the licensing of amateur radio operators in the United States. Here are some of the key licensing requirements for amateur radio operators:

1. Passing the FCC Exam - To obtain a ham radio license, you must pass an exam administered by the FCC. The exam covers basic electronics, radio wave propagation, operating practices, and FCC regulations.
2. License Classes - There are three classes of ham radio licenses: Technician, General, and Extra. The Technician class license allows operators to use VHF and UHF frequencies, while the General and Extra class licenses allow access to a wider range of frequencies.
3. Renewal - Ham radio licenses must be renewed every ten years. Renewal requires passing an exam or providing proof of continuing education in radio communications.
4. Call Sign - Each licensed amateur radio operator is assigned a unique call sign by the FCC. The call sign is used to identify the operator and the location of the station.
5. Operating Restrictions - There are certain operating restrictions that apply to ham radio operators, including restrictions on power output and antenna height.

It's important for ham radio operators to understand and comply with these licensing requirements to ensure that they can legally operate on the amateur radio frequencies. Failure to comply with these requirements can result in fines, license revocation, and legal penalties.

Additionally, the licensing process provides operators with a basic understanding of radio communications, including basic electronics and regulations. This knowledge is important for operating ham radio equipment safely and effectively, and for communicating with other operators during emergency situations.

Passing the FCC Exam

To obtain a ham radio license, you must pass an exam administered by the Federal Communications Commission (FCC). The exam is designed to test your knowledge of basic electronics, radio wave propagation, operating practices, and FCC regulations. Here are some key things to know about the FCC exam:

1. Exam Elements - There are three exam elements: Technician (Element 2), General (Element 3), and Extra (Element 4). Each exam covers more advanced topics than the previous one, and passing a higher-level exam provides access to more operating privileges.
2. Exam Questions - The exam consists of multiple-choice questions, and the number of questions and time allowed for each exam element varies. The Technician exam has 35 questions, the General exam has 35 questions, and the Extra exam has 50 questions.
3. Passing Score - To pass the exam, you must achieve a minimum score of 74% on each exam element. If you fail an exam element, you can retake it after a waiting period of at least 14 days.
4. Exam Fees - There is a fee to take the FCC exam, which varies depending on the exam location and the organization administering the exam.
5. Study Materials - There are many study materials available to help prepare for the FCC exam, including study

guides, practice exams, and online courses.

Passing the FCC exam is an important step in obtaining a ham radio license, as it demonstrates your knowledge of basic electronics, radio wave propagation, operating practices, and FCC regulations. With a license, you can legally operate on amateur radio frequencies and communicate with other ham radio operators around the world.

License Classes

There are three classes of ham radio licenses in the United States, as established by the Federal Communications Commission (FCC). The classes are Technician, General, and Extra.

1. Technician Class - The Technician class license is the entry-level license and allows access to all amateur radio frequencies above 30 MHz. This includes the popular VHF and UHF bands, which are used for local communication and some limited long-distance communication.
2. General Class - The General class license allows access to a wider range of amateur radio frequencies, including the high frequency (HF) bands that are used for long-distance communication. General class license holders can operate on all amateur radio frequencies above 50 MHz.
3. Extra Class - The Extra class license is the highest level of ham radio license and provides access to all amateur radio frequencies. Extra class license holders have the widest range of operating privileges and can communicate on all amateur radio bands, including the HF bands that are used for long-distance communication.

Each class of license requires passing a specific FCC exam that covers basic electronics, radio wave propagation, operating practices, and FCC regulations. Each exam is more difficult than the previous one, and covers more advanced topics.

As ham radio operators progress through the license classes, they gain access to more frequencies and operating privileges. This allows them to communicate with other operators over longer distances and on more advanced modes of communication, such as digital modes and satellite communication.

Technician Class

The Technician class ham radio license is the entry-level license and allows access to all amateur radio frequencies above 30 MHz. Here are some key things to know about the Technician class license:

1. Exam Element - The Technician exam (Element 2) consists of 35 multiple-choice questions that cover basic electronics, radio wave propagation, operating practices, and FCC regulations.
2. Operating Privileges - With a Technician class license, you can communicate using voice, Morse code, and digital modes on all amateur radio frequencies above 30 MHz. This includes the VHF and UHF bands, which are used for local communication and some limited long-distance communication.
3. Call Sign - Each licensed amateur radio operator is assigned a unique call sign by the FCC. Technician class license holders are assigned call signs that start with the letter "K," "N," or "W."

4. License Renewal - Ham radio licenses must be renewed every ten years. Technician class license holders can renew their licenses by passing an exam or providing proof of continuing education in radio communications.
5. Upgrading - Technician class license holders can upgrade to the General or Extra class licenses by passing the corresponding FCC exams. Upgrading allows access to more operating privileges and frequencies.

The Technician class license is a great way to get started in ham radio and provides access to a wide range of operating privileges on the VHF and UHF bands. With a license, you can communicate with other ham radio operators in your local area and participate in emergency communications during disasters and other events.

General Class

The General class ham radio license is the second level of license and provides access to a wider range of amateur radio frequencies, including the high frequency (HF) bands that are used for long-distance communication. Here are some key things to know about the General class license:

1. Exam Element - The General exam (Element 3) consists of 35 multiple-choice questions that cover more advanced topics in electronics, radio wave propagation, operating practices, and FCC regulations than the Technician exam.
2. Operating Privileges - With a General class license, you can communicate using voice, Morse code, and digital modes on all amateur radio frequencies above 50 MHz.

This includes access to the HF bands, which are used for long-distance communication around the world.

3. Call Sign - Each licensed amateur radio operator is assigned a unique call sign by the FCC. General class license holders are assigned call signs that start with the letter "K," "N," or "W."
4. License Renewal - Ham radio licenses must be renewed every ten years. General class license holders can renew their licenses by passing an exam or providing proof of continuing education in radio communications.
5. Upgrading - General class license holders can upgrade to the Extra class license by passing the corresponding FCC exam. Upgrading allows access to all amateur radio frequencies and operating privileges.

The General class license is a great way to expand your ham radio operating privileges and communicate over long distances. With a license, you can participate in emergency communications during disasters and other events, and communicate with other ham radio operators around the world.

Extra Class

The Extra class ham radio license is the highest level of ham radio license and provides access to all amateur radio frequencies. Here are some key things to know about the Extra class license:

1. Exam Element - The Extra exam (Element 4) consists of 50 multiple-choice questions that cover the most advanced topics in electronics, radio wave propagation, operating practices, and FCC regulations.

2. Operating Privileges - With an Extra class license, you can communicate using voice, Morse code, and digital modes on all amateur radio frequencies. This includes access to all HF bands, as well as additional frequencies in other bands.
3. Call Sign - Each licensed amateur radio operator is assigned a unique call sign by the FCC. Extra class license holders are assigned call signs that start with the letter "K," "N," or "W."
4. License Renewal - Ham radio licenses must be renewed every ten years. Extra class license holders can renew their licenses by passing an exam or providing proof of continuing education in radio communications.
5. Continuing Education - Extra class license holders are encouraged to continue their education in radio communications and stay up-to-date with the latest advancements in the field.

The Extra class license provides the widest range of operating privileges and frequencies, allowing you to communicate with other ham radio operators around the world. With a license, you can participate in emergency communications during disasters and other events, and use advanced communication modes like digital modes and satellite communication.

Exam Preparation

Preparing for the ham radio license exam requires dedication and effort. Here are some steps you can take to prepare for the exam:

1. Choose Your Study Materials - There are many study materials available, including books, online courses, and practice exams. Choose the materials that work best for you based on your learning style and experience.
2. Set a Study Schedule - Create a study schedule that works for your schedule and stick to it. Set specific goals and deadlines for each study session.
3. Take Practice Exams - Practice exams are a great way to prepare for the exam and identify areas where you may need more study. There are many online resources that offer free practice exams for all license classes.
4. Join a Study Group - Joining a study group is a great way to connect with other aspiring ham radio operators and share knowledge and resources. You can find study groups online or through local ham radio clubs.
5. Understand the Exam Format - The ham radio license exam consists of multiple-choice questions and covers a variety of topics. Each license class has a different number of exam elements. Make sure you understand the exam format and what topics will be covered.
6. Review the Exam Question Pool - The FCC publishes a question pool for each exam element. Review the question pool and make sure you understand the concepts and topics covered.
7. Get Hands-On Experience - Practice using ham radio equipment and operating on the amateur radio frequencies. This will help you understand the concepts covered on the exam and build confidence.

Preparing for the ham radio license exam takes time and effort, but the rewards of obtaining your license are well worth it.

With a license, you can legally operate on the amateur radio frequencies and communicate with other ham radio operators around the world. You can also participate in emergency communications during disasters and other events.

License Cost

In the United States, there is no fee to obtain a ham radio license from the Federal Communications Commission (FCC). However, there are fees associated with the exam and license renewal process. Here are some key things to know about ham radio license fees:

1. Exam Fees - There is a fee to take the FCC exam, which varies depending on the exam location and the organization administering the exam. The fees typically range from $10 to $15 per exam element.
2. License Renewal Fees - There is a fee to renew your ham radio license, which varies depending on the type of license and the organization administering the exam. The fees typically range from $10 to $15.
3. Vanity Call Sign Fees - If you want to obtain a vanity call sign (a call sign that you choose yourself), there is a fee to do so. The fee is currently $21.40 and is non-refundable.
4. Volunteer Examiner Fees - Volunteer examiners are individuals who administer the FCC exam. Some organizations charge a fee for their services, which is typically between $5 and $15 per exam element.
5. Other Fees - There may be other fees associated with ham radio, such as equipment costs, club memberships, and event fees.

Overall, the fees associated with obtaining and maintaining a ham radio license are relatively low compared to other hobbies and interests. With a license, you can legally operate on the amateur radio frequencies and communicate with other ham radio operators around the world.

Renewal

Ham radio licenses must be renewed every ten years to remain valid. Here are some key things to know about license renewal:

1. Exam or Continuing Education - To renew your license, you must either pass an exam or provide proof of continuing education in radio communications. The FCC offers exam sessions at various locations throughout the country, and there are many resources available to help you prepare for the exam.
2. License Expiration - If you do not renew your license before it expires, your call sign will be removed from the FCC database and you will no longer be able to legally operate on the amateur radio frequencies.
3. Grace Period - The FCC allows a two-year grace period for license renewal. This means that you can renew your license up to two years after it expires without having to retake the exam.
4. Vanity Call Sign - If you have a vanity call sign (a call sign that you chose yourself), you will need to renew it separately from your license.
5. License Renewal Fees - There is a fee to renew your ham radio license, which varies depending on the type of license and the organization administering the exam.

Renewing your ham radio license is an important step in maintaining your ability to operate on the amateur radio frequencies. By staying up-to-date with the latest regulations and best practices, you can continue to communicate with other ham radio operators and participate in emergency communications during disasters and other events.

Call Sign

A call sign is a unique combination of letters and numbers that identifies a licensed radio station or operator. In the United States, call signs for ham radio operators are assigned by the Federal Communications Commission (FCC) and consist of one or two letters, followed by a number, and then by one or two more letters. Here are some key things to know about call signs for ham radio operators:

1. FCC Assignment - Call signs for ham radio operators in the United States are assigned by the FCC. The FCC uses an automated system to assign call signs based on the operator's location and license class.
2. License Class - The call sign prefix (the first letter or letters of the call sign) indicates the license class of the operator. For example, call signs starting with "K" are assigned to operators with Technician, General, or Extra class licenses.
3. Vanity Call Signs - Ham radio operators can apply for a vanity call sign, which is a call sign that they choose themselves. Vanity call signs typically have a special meaning to the operator or are easier to remember than their original call sign. There is a fee to obtain a vanity call

sign.

4. Identification - Ham radio operators are required to identify their station at least once every 10 minutes during a communication and at the end of a communication. This is typically done by speaking the call sign using voice or Morse code.
5. Call Sign Lookup - There are many websites and databases that allow you to look up call signs for ham radio operators around the world. This can be a useful tool for finding and contacting other operators.

Your call sign is a unique identifier that represents you as a licensed ham radio operator. By using your call sign during communications, you can build a reputation and make connections with other operators around the world.

Operating Restrictions

Ham radio operators in the United States are subject to certain operating restrictions, which are put in place to protect other radio services and ensure fair use of the amateur radio frequencies. Here are some key things to know about operating restrictions for ham radio operators:

1. Frequency Limits - Ham radio operators are only allowed to operate on certain frequencies, which are allocated specifically for amateur radio use. These frequencies vary depending on the operator's license class and the type of communication they are engaging in.
2. Power Limits - Ham radio operators are limited in the amount of power they can use to transmit their signal.

The maximum power output varies depending on the operator's license class and the frequency they are using.

3. Interference - Ham radio operators are responsible for ensuring that their transmissions do not interfere with other radio services or communications. They are also responsible for detecting and reporting any interference that they experience.
4. Obscenity and Profanity - Ham radio operators are not allowed to use obscene or profane language during communications.
5. Emergency Communications - Ham radio operators are allowed to engage in emergency communications during disasters and other events, even if they are not licensed for the frequency they are using.
6. Third-Party Communications - Ham radio operators are allowed to communicate with third parties, such as non-licensed individuals, as long as they are present and participating in the communication.
7. Broadcasts - Ham radio operators are not allowed to transmit broadcasts or music, except in certain limited circumstances.

By following these operating restrictions, ham radio operators can ensure fair use of the amateur radio frequencies and avoid interference with other radio services. They can also participate in emergency communications and use their skills and equipment to help others during disasters and other events.

Equipment Considerations

When it comes to ham radio, having the right equipment is essential for successful communication. Here are some key equipment considerations for ham radio operators:

1. Transceiver - The transceiver is the main piece of equipment that allows you to transmit and receive radio signals. It is important to choose a transceiver that meets your needs and is compatible with the frequencies and modes you want to use.
2. Antenna - The antenna is a crucial component of your ham radio station. It is responsible for transmitting and receiving radio signals. There are many types of antennas available, including wire antennas, vertical antennas, and directional antennas. Choose an antenna that is compatible with your transceiver and the frequencies you want to use.
3. Power Source - Your ham radio station will need a reliable power source. This can be a battery, a generator, or a power supply. Make sure you have backup power sources in case of an emergency.
4. Accessories - There are many accessories that can enhance your ham radio experience, including headphones, microphones, and filters. Consider what accessories would be helpful for your specific needs.
5. Mobile or Portable Equipment - If you plan to operate your ham radio station while on the go, you will need mobile or portable equipment. This can include handheld transceivers, mobile antennas, and battery packs.
6. Test Equipment - It is important to have test equipment

to ensure that your ham radio station is working properly. This can include an antenna analyzer, a SWR meter, and a multimeter.

7. Computer and Software - Many ham radio operators use a computer and software to log contacts and manage their station. Choose software that is compatible with your transceiver and operating system.

When choosing ham radio equipment, it is important to consider your specific needs and budget. It may be helpful to consult with other ham radio operators or visit a ham radio store for advice and recommendations.

Transceiver

There are several types of radios available for ham radio operators. Here are some of the most common types:

1. Handheld Radios - Handheld radios are portable and can be easily carried with you. They are often used for short-range communications and are ideal for use in the field or during emergencies.
2. Mobile Radios - Mobile radios are designed to be installed in a vehicle or at a fixed location. They are more powerful than handheld radios and can be used for longer-range communications.
3. Base Station Radios - Base station radios are typically used in a fixed location and are more powerful than mobile or handheld radios. They are often used for long-range communications and can be connected to external antennas for improved performance.

4. Portable Radios - Portable radios are similar to handheld radios, but are designed to be more rugged and durable. They are often used by hikers, campers, and other outdoor enthusiasts.
5. SDR Radios - Software-defined radios (SDR) are a type of radio that uses software to define the frequency range and modulation scheme. They are highly flexible and can be used for a wide range of applications.
6. QRP Radios - QRP (low power) radios are designed to operate with low power levels, typically less than 5 watts. They are often used by amateur radio enthusiasts who enjoy the challenge of communicating with minimal power.

When choosing a radio, it is important to consider your specific needs and the type of communication you will be engaging in. It is also important to ensure that your radio is compatible with the frequencies and modes you want to use.

Antenna

An antenna is a crucial component of a ham radio station. It is responsible for transmitting and receiving radio signals. Here are some key considerations when choosing an antenna:

1. Frequency Range - Antennas are designed to operate within specific frequency ranges. It is important to choose an antenna that is compatible with the frequencies you want to use.
2. Gain - Antennas can have varying levels of gain, which refers to the ability of the antenna to direct and amplify signals in a particular direction. Higher-gain antennas

can improve the range and quality of your communication.

3. Polarization - Antennas can be polarized in different ways, such as horizontal or vertical. It is important to choose an antenna that is compatible with the polarization of the signals you want to transmit and receive.
4. Directionality - Some antennas are designed to be directional, meaning they transmit and receive signals in a particular direction. This can be useful for long-range communications and reducing interference.
5. Size and Installation - Antennas come in a variety of sizes and can be installed in different ways, such as on a roof, a mast, or a portable mount. Consider the size and installation requirements of the antenna when choosing one for your ham radio station.
6. Cost - Antennas can vary greatly in price, from less than $50 to thousands of dollars. Consider your budget when choosing an antenna, but also keep in mind that a higher-quality antenna can improve the performance of your ham radio station.
7. Accessories - Some antennas may require accessories, such as coaxial cable, connectors, and baluns. Make sure you have the necessary accessories for your antenna to function properly.

When choosing an antenna, it is important to consider your specific needs and the type of communication you will be engaging in. It may be helpful to consult with other ham radio operators or visit a ham radio store for advice and recommendations.

Power Source

A reliable power source is crucial for any ham radio station. Here are some key considerations when choosing a power source:

1. Battery - Batteries are a popular choice for portable and emergency communications. They can be recharged using a variety of methods, such as solar panels, generators, or a power supply.
2. Generator - Generators are a reliable source of power for ham radio stations that require a lot of power or need to operate for an extended period of time. They can be fueled by gasoline, diesel, or propane.
3. Power Supply - A power supply is a device that converts AC power from a wall outlet to DC power that can be used by your ham radio station. They are typically used for base station radios that are located in a fixed location.
4. Solar Power - Solar power is a renewable source of energy that can be used to charge batteries or power small ham radio stations. Solar panels can be portable or fixed, depending on your needs.
5. Backup Power - It is important to have backup power sources in case of an emergency or power outage. This can include a backup battery, a generator, or a power supply.

When choosing a power source, it is important to consider your specific needs and the type of communication you will be engaging in. It may be helpful to consult with other ham radio operators or visit a ham radio store for advice and recommendations.

Accessories

There are many accessories available for ham radio operators that can enhance your communication experience. Here are some common accessories:

1. Headphones - Headphones are useful for reducing background noise and allowing you to hear signals more clearly.
2. Microphone - A microphone is used to transmit your voice during communication. There are many types of microphones available, including handheld, desk-mounted, and headset.
3. Filters - Filters can help reduce interference and improve the quality of your communication. They can be used to filter out unwanted signals, such as noise from power lines or other electronic devices.
4. Amplifiers - Amplifiers can boost the signal strength of your ham radio station, improving your communication range and quality.
5. Tuners - A tuner is used to match the impedance of your antenna to the impedance of your transceiver, improving the efficiency of your antenna system.
6. Rotators - Rotators are used to rotate directional antennas, allowing you to direct your antenna towards a specific location or signal.
7. SWR Meters - SWR meters are used to measure the standing wave ratio of your antenna system, ensuring that your system is working efficiently.
8. Antenna Analyzers - Antenna analyzers are used to measure the performance of your antenna system, including

impedance, resonance, and SWR.

When choosing accessories, it is important to consider your specific needs and the type of communication you will be engaging in. It may be helpful to consult with other ham radio operators or visit a ham radio store for advice and recommendations.

Mobile or Portable Equipment

Mobile or portable equipment is ideal for ham radio operators who need to communicate while on the move. Here are some common types of mobile or portable equipment:

1. Handheld Radios - Handheld radios are small, portable devices that can be easily carried with you. They are ideal for short-range communications and are often used in the field or during emergencies.
2. Mobile Radios - Mobile radios are designed to be installed in a vehicle or at a fixed location. They are more powerful than handheld radios and can be used for longer-range communications.
3. Portable Antennas - Portable antennas are designed to be lightweight and easy to transport. They can be set up quickly and are ideal for use in the field or during emergencies.
4. Battery Packs - Battery packs are used to power mobile or portable equipment. They can be recharged using a variety of methods, such as solar panels or a power supply.
5. Portable Amplifiers - Portable amplifiers can be used to boost the signal strength of your mobile or portable equipment, improving your communication range and

quality.

When choosing mobile or portable equipment, it is important to consider your specific needs and the type of communication you will be engaging in. It may be helpful to consult with other ham radio operators or visit a ham radio store for advice and recommendations.

Test Equipment

Test equipment is used by ham radio operators to measure and test their equipment to ensure that it is working properly. Here are some common types of test equipment:

1. Multimeter - A multimeter is a device used to measure voltage, current, and resistance. It is useful for testing and troubleshooting electrical circuits.
2. SWR Meter - An SWR meter is used to measure the standing wave ratio of your antenna system, ensuring that your system is working efficiently.
3. Antenna Analyzer - An antenna analyzer is used to measure the performance of your antenna system, including impedance, resonance, and SWR.
4. Signal Generator - A signal generator is used to generate a specific signal, which can be used to test and troubleshoot equipment.
5. Oscilloscope - An oscilloscope is a device used to measure and display waveforms. It is useful for testing and troubleshooting electronic circuits.
6. Spectrum Analyzer - A spectrum analyzer is used to display and analyze frequency spectra of signals. It is

useful for detecting interference and troubleshooting electronic circuits.

When choosing test equipment, it is important to consider your specific needs and the type of equipment you will be testing. It may be helpful to consult with other ham radio operators or visit a ham radio store for advice and recommendations.

Computer and Software

Computers and software can be useful tools for ham radio operators. Here are some common uses of computers and software in ham radio:

1. Logging Software - Logging software can be used to log and organize contacts, making it easier to keep track of your communication activity.
2. Digital Modes Software - Digital modes software can be used to send and receive digital signals, which can be more efficient than voice communication.
3. Antenna Design Software - Antenna design software can be used to design and analyze antenna systems, helping you to optimize your antenna for your specific needs.
4. Propagation Prediction Software - Propagation prediction software can be used to predict radio propagation, allowing you to plan your communication based on the expected conditions.
5. Contesting Software - Contesting software can be used to participate in ham radio contests, which involve making as many contacts as possible within a set time period.
6. Remote Control Software - Remote control software can

be used to control your ham radio station from a remote location, allowing you to operate your station while away from home.

7. SDR Software - SDR (Software-Defined Radio) software can be used to receive and decode signals using a computer and a radio receiver.

When choosing computer and software for ham radio, it is important to consider your specific needs and the type of communication you will be engaging in. It may be helpful to consult with other ham radio operators or visit a ham radio store for advice and recommendations.

Basic Operating Procedure

Once you have your ham radio equipment set up and have obtained your license, it's important to understand basic operating procedures. Here are some key procedures to keep in mind:

1. Listen Before Transmitting - Before transmitting, it is important to listen to the frequency to ensure that it is clear and that you are not interfering with other ongoing communication.
2. Identify Yourself - When you begin transmitting, identify yourself with your call sign. It is also important to identify yourself periodically throughout your transmission.
3. Speak Clearly - When transmitting, speak clearly and at a moderate pace. This will ensure that your message is clearly understood by other operators.
4. Use Standard Phrases - Ham radio operators use standard

phrases and abbreviations to ensure efficient and clear communication. It's important to learn and use these phrases to communicate effectively.

5. Follow Protocols - Ham radio operators follow protocols to ensure that communication is efficient and respectful. It's important to follow these protocols to avoid confusion and misunderstandings.
6. Be Patient - Ham radio communication can be slow and sometimes interrupted by interference or other factors. It's important to be patient and wait for other operators to finish before transmitting.
7. Monitor Your Equipment - It's important to monitor your equipment while communicating to ensure that it is working properly and that you are transmitting and receiving effectively.
8. End Your Transmission - When you have finished transmitting, end your transmission with a standard phrase such as "over" or "clear". This will let other operators know that you are finished speaking and they can respond.

By following these basic operating procedures, you can communicate effectively and efficiently on ham radio frequencies.

Frequencies and modes

Ham radio operators have access to a wide range of frequencies and modes for communication. Here are some common frequencies and modes used in ham radio:

1. HF (High Frequency) - HF frequencies are used for long-range communication and can travel long distances

through the ionosphere. HF frequencies range from 3 to 30 MHz.

2. VHF (Very High Frequency) - VHF frequencies are used for short to medium range communication and are often used for local communication or during emergencies. VHF frequencies range from 30 to 300 MHz.
3. UHF (Ultra High Frequency) - UHF frequencies are also used for short to medium range communication and are often used for satellite and repeater communication. UHF frequencies range from 300 to 3000 MHz.
4. AM (Amplitude Modulation) - AM is a mode of communication that uses changes in amplitude to transmit voice and other signals. It is commonly used on HF frequencies.
5. SSB (Single Sideband) - SSB is a mode of communication that uses a narrower bandwidth than AM, making it more efficient for long-range communication on HF frequencies.
6. FM (Frequency Modulation) - FM is a mode of communication that uses changes in frequency to transmit voice and other signals. It is commonly used on VHF and UHF frequencies.
7. Digital Modes - Digital modes are modes of communication that use digital signals to transmit data, images, or voice. Examples of digital modes include RTTY, PSK31, and JT65.

When choosing frequencies and modes for communication, it is important to consider your specific needs and the distance and location of the other operator(s). It may be helpful to consult with other ham radio operators or use online resources to determine the best frequency and mode for your communication.

Making contacts

Making contacts is a fundamental part of ham radio communication. Here are some tips for making successful contacts:

1. Listen First - Before transmitting, listen to the frequency to ensure that it is clear and that you are not interfering with other ongoing communication.
2. Choose the Right Frequency and Mode - Choose a frequency and mode that is appropriate for the distance and location of the other operator(s). Use online resources or consult with other ham radio operators to determine the best frequency and mode for your communication.
3. Identify Yourself - When you begin transmitting, identify yourself with your call sign. It is also important to identify yourself periodically throughout your transmission.
4. Be Clear and Concise - When transmitting, speak clearly and concisely. Use standard phrases and avoid unnecessary words or chatter.
5. Be Patient - Making contacts on ham radio can take time, especially if you are trying to contact operators in other countries or distant locations. Be patient and persistent.
6. Use Proper Etiquette - Use proper etiquette when communicating on ham radio, including following protocols and being respectful to other operators.
7. Record Your Contacts - Record your contacts in a logbook to keep track of your communication activity and to ensure that you are meeting licensing requirements.

By following these tips, you can make successful contacts on ham radio frequencies and enjoy the many benefits of ham radio communication.

5

Setting Up Your Station

Amateur Radio Base Station

Setting up your ham radio station is a critical step in preparing for emergency communications. Here are some key considerations when setting up your station:

1. Choosing a Location - Choose a location for your station that is free from interference and has access to power and a reliable antenna support system.
2. Power Sources - Choose a reliable power source for your station, such as a battery or generator, and ensure that you have backup power available in case of a power outage.
3. Antennas - Choose an appropriate antenna for your station, taking into consideration the frequency and mode of communication you will be using, as well as the location of your station.
4. Communication Modes - Choose the communication modes that are appropriate for your station setup and communication needs, such as SSB, FM, or digital modes.
5. Radio Equipment - Choose radio equipment that is appropriate for your communication needs, such as a transceiver, amplifier, or tuner.
6. Accessories - Choose accessories that will help you to operate your station efficiently and effectively, such as a microphone, headphones, and a key or paddle for CW (Morse code) communication.
7. Station Grounding - Proper station grounding is important for safety and for preventing interference. Ensure that your station is properly grounded according to local regulations.
8. Interference Reduction - Take steps to reduce interference from other electronic devices, such as using filters and ferrite chokes.

By considering these factors when setting up your station, you can ensure that your station is ready for effective emergency communication.

Power Source

Choosing a reliable power source is an important consideration when setting up your ham radio station for emergency communication. Here are some common power sources for ham radio stations:

1. AC Power - AC power from a wall outlet is a common source of power for ham radio stations, but it is not always reliable during emergencies.
2. DC Power Supply - A DC power supply can provide reliable power to your ham radio station, but it requires a 12-volt DC power source.
3. Batteries - Batteries can provide portable power for your ham radio station, but they require regular charging or replacement.
4. Generators - Generators can provide reliable backup power during extended power outages, but they can be noisy and require fuel.

When choosing a power source for your ham radio station, consider factors such as reliability, portability, and noise level. It may be helpful to have multiple power sources available, including backup power sources in case of a power outage. Additionally, it is important to follow safety guidelines when working with electrical equipment, including grounding your station and using proper fuses and circuit breakers.

Batteries

Batteries are a common power source for portable and emergency ham radio stations. Here are some considerations when using batteries for your ham radio station:

1. Type of Battery - Choose a battery type that is appropriate for your station's power needs, such as lead-acid, nickel-cadmium, or lithium-ion batteries.
2. Capacity - Choose a battery with sufficient capacity to power your station for the desired duration of operation. Consider factors such as the power consumption of your radio equipment and the frequency of use.
3. Charging - Batteries require regular charging to maintain their capacity. Use a battery charger that is appropriate for the type of battery you are using, and follow the manufacturer's instructions for charging.
4. Battery Maintenance - Proper maintenance of your batteries can extend their lifespan and ensure reliable operation. Keep batteries clean and free from corrosion, and avoid overcharging or discharging them.
5. Backup Batteries - Have backup batteries available in case of a power outage or if your primary batteries are depleted. This will ensure that you can continue operating your ham radio station during emergencies.

By considering these factors when using batteries for your ham radio station, you can ensure reliable power during emergency communications.

Generators

Generators are a reliable source of backup power for ham radio stations during extended power outages. Here are some considerations when using generators for your ham radio station:

1. Type of Generator - Choose a generator that is appropriate for your station's power needs, such as a portable gasoline or diesel generator, or a propane or natural gas standby generator.
2. Power Output - Choose a generator with sufficient power output to meet the power requirements of your station's equipment, including any additional equipment that may be needed during emergency operations.
3. Noise Level - Generators can be noisy, so consider the noise level of the generator you are using, especially if you are operating in a residential area.
4. Fuel Storage - Generators require fuel to operate, so ensure that you have sufficient fuel storage and that you follow safe fuel handling practices.
5. Maintenance - Proper maintenance of your generator is important to ensure reliable operation during emergencies. Follow the manufacturer's instructions for maintenance and keep your generator in good working order.
6. Safety - When using a generator, follow proper safety guidelines, such as keeping the generator away from living areas to prevent carbon monoxide poisoning, and ensuring that it is properly grounded.

By considering these factors when using generators for your

ham radio station, you can ensure reliable backup power during emergency communications.

A natural gas generator is a type of standby generator that runs on natural gas. Here are some considerations when using a natural gas generator for your ham radio station:

1. Power Output - Choose a natural gas generator with sufficient power output to meet the power requirements of your station's equipment, including any additional equipment that may be needed during emergency operations.
2. Fuel Supply - Natural gas generators require a reliable natural gas supply, which may not be available during natural disasters or other emergencies. Ensure that you have a backup fuel supply, such as propane or gasoline, available in case of a natural gas outage.
3. Noise Level - Natural gas generators are generally quieter than gasoline or diesel generators, but they still produce some noise. Consider the noise level of the generator you are using, especially if you are operating in a residential area.
4. Maintenance - Proper maintenance of your natural gas generator is important to ensure reliable operation during emergencies. Follow the manufacturer's instructions for maintenance and keep your generator in good working order.
5. Safety - When using a natural gas generator, follow proper safety guidelines, such as ensuring that it is properly ventilated to prevent carbon monoxide poisoning, and ensuring that it is properly grounded.

By considering these factors when using a natural gas generator for your ham radio station, you can ensure reliable backup power during emergency communications.

Antenna

Antennas are a critical component of a ham radio station, as they allow for the transmission and reception of radio signals. Here are some considerations when choosing and setting up your antennas for emergency communication:

1. Frequency - Choose an antenna that is appropriate for the frequency range that you will be using for communication. Different types of antennas are optimized for different frequency ranges, such as vertical antennas for high-frequency communication and Yagi antennas for VHF/UHF communication.
2. Location - Choose a location for your antenna that is free from obstructions and interference. An outdoor location is preferred for most antennas, as this allows for maximum signal strength and range.
3. Height - The height of your antenna can have a significant impact on signal strength and range. Higher antennas generally provide better performance, but may require additional support and may be subject to regulations and local zoning restrictions.
4. Grounding - Proper grounding is important for safety and for minimizing interference. Ensure that your antenna is properly grounded according to local regulations.
5. Type of Antenna - Choose an antenna type that is appropriate for your communication needs, such as a vertical,

dipole, Yagi, or wire antenna.

6. Antenna Tuner - An antenna tuner can help match your antenna to your transmitter and improve the efficiency of your communication.
7. Antenna Maintenance - Proper maintenance of your antenna is important for reliable performance. Keep your antenna clean and free from corrosion, and ensure that it is properly supported and secured.

By considering these factors when choosing and setting up your antennas, you can ensure reliable communication during emergency situations.

Antenna Frequency

The frequency of an antenna is an important consideration when setting up your ham radio station for emergency communication. Here are some factors to consider:

1. Resonant Frequency - An antenna's resonant frequency is the frequency at which it is most efficient in transmitting and receiving signals. It is important to choose an antenna that is resonant at the frequency you will be using for communication.
2. Frequency Range - Different types of antennas are optimized for different frequency ranges, such as vertical antennas for high-frequency communication and Yagi antennas for VHF/UHF communication. Ensure that your antenna is appropriate for the frequency range you will be using.
3. Bandwidth - The bandwidth of an antenna refers to the

range of frequencies over which it can effectively transmit and receive signals. A wider bandwidth is generally better for emergency communication, as it allows for more flexibility in frequency selection.

4. Radiation Pattern - The radiation pattern of an antenna refers to the direction and strength of the signals it transmits and receives. Choose an antenna with a radiation pattern that is appropriate for your communication needs.
5. Polarization - The polarization of an antenna refers to the orientation of the electric and magnetic fields of the signal it transmits and receives. Choose an antenna with a polarization that is appropriate for the polarization of the signals you will be transmitting and receiving.

By considering these factors when choosing an antenna for emergency communication, you can ensure efficient and reliable communication during emergency situations.

Antenna Location

The location of your antenna is an important consideration when setting up your ham radio station for emergency communication. Here are some factors to consider:

1. Obstructions - Choose a location for your antenna that is free from obstructions such as buildings, trees, or power lines. Obstructions can block or weaken the signal, reducing the range and quality of your communication.
2. Interference - Choose a location that is free from sources of interference such as electrical equipment or nearby radio towers. Interference can disrupt your communication

and make it difficult to hear or be heard.

3. Height - The height of your antenna can have a significant impact on signal strength and range. Higher antennas generally provide better performance, but may require additional support and may be subject to regulations and local zoning restrictions.
4. Grounding - Proper grounding is important for safety and for minimizing interference. Ensure that your antenna is properly grounded according to local regulations.
5. Accessibility - Choose a location that is easily accessible for installation, maintenance, and repairs. If your antenna is located in a difficult-to-reach location, it may be difficult to maintain or repair in case of damage.

By considering these factors when choosing a location for your antenna, you can ensure efficient and reliable communication during emergency situations.

Antenna Height

The height of your antenna is an important consideration when setting up your ham radio station for emergency communication. Here are some factors to consider:

1. Line-of-Sight - The higher your antenna is, the greater the line-of-sight range will be. This can be especially important during emergency situations when you need to communicate over long distances.
2. Obstructions - A higher antenna may allow you to avoid obstructions such as buildings, trees, or power lines, which can block or weaken the signal.

3. Regulations - Depending on where you live, there may be regulations or zoning restrictions that limit the height of your antenna. It is important to research these regulations before setting up your antenna.
4. Stability - A taller antenna may require additional support to ensure stability during high winds or other extreme weather conditions. Consider the strength of the mounting structure and guy wires that will be needed to support the antenna at greater heights.
5. Cost - A taller antenna may require more expensive equipment and installation costs than a shorter one.

By considering these factors when choosing the height of your antenna, you can ensure efficient and reliable communication during emergency situations.

Antenna Grounding

Proper grounding of your antenna is an important consideration when setting up your ham radio station for emergency communication. Here are some factors to consider:

1. Safety - Grounding your antenna provides a safe path for excess electrical energy to discharge in case of a lightning strike or other electrical surge. It can protect your equipment and prevent damage or injury.
2. Interference - Proper grounding can minimize interference from nearby electrical equipment or power lines. It can improve the signal quality and reduce background noise.
3. Regulations - Depending on where you live, there may

be regulations or zoning restrictions that require proper grounding of your antenna. It is important to research these regulations before setting up your antenna.

4. Soil conductivity - The conductivity of the soil around your antenna can affect the effectiveness of the grounding system. It is important to use proper grounding rods and ensure they are driven deep enough into the soil.
5. Grounding system - Proper grounding of your antenna requires a good grounding system, which includes grounding rods, conductors, and connectors. Ensure that you use the proper materials and follow proper installation procedures.

By considering these factors when grounding your antenna, you can ensure efficient and reliable communication during emergency situations.

Antenna Type

There are several types of antennas that can be used for ham radio communication. Here are some of the most common types:

1. Dipole antenna - A dipole antenna is a simple wire antenna that consists of two parallel conductors. It is a balanced antenna that is resonant at a specific frequency.
2. Vertical antenna - A vertical antenna is a type of monopole antenna that is vertically polarized. It is typically used for frequencies below 30 MHz.
3. Yagi antenna - A Yagi antenna is a directional antenna that consists of a driven element and several parasitic elements.

It is used for VHF and UHF frequencies and provides high gain and directivity.

4. Loop antenna - A loop antenna is a type of directional antenna that consists of a loop of wire. It is used for frequencies below 30 MHz and can be either horizontally or vertically polarized.
5. Log-periodic antenna - A log-periodic antenna is a type of directional antenna that is designed to operate over a wide range of frequencies. It consists of a series of elements of varying length and spacing.
6. Ground-plane antenna - A ground-plane antenna is a type of vertical antenna that uses the ground as a reflector to enhance the signal.
7. Discone antenna - A discone antenna is a wideband antenna that consists of a cone-shaped radiator and a disc-shaped ground plane.

Each type of antenna has its own strengths and weaknesses, and the best choice will depend on the frequency range, polarization, and directional requirements of your communication needs.

Antenna Tuner

An antenna tuner is a device that can adjust the impedance of the antenna system to match the transmitter's output impedance. This can improve the efficiency of the antenna system and help to reduce interference.

Here are some factors to consider when using an antenna tuner for ham radio communication:

1. Frequency range - Choose an antenna tuner that is compatible with the frequency range of your transmitter and antenna system.
2. Power handling capacity - Choose an antenna tuner that can handle the power output of your transmitter.
3. Impedance matching range - Choose an antenna tuner that has a matching range that is appropriate for your antenna system.
4. Type of tuner - There are several types of antenna tuners, including manual, automatic, and remote-controlled. Choose the type that best fits your needs and operating preferences.
5. Installation - Install the antenna tuner close to the antenna to minimize transmission line losses.
6. Operating procedure - Follow the manufacturer's instructions and recommended operating procedures for your specific antenna tuner.

An antenna tuner can be a useful tool for optimizing the efficiency of your antenna system and improving the quality of your ham radio communication.

Antenna Maintenance

Proper maintenance of your antenna is important to ensure optimal performance and reliability. Here are some tips for maintaining your antenna:

1. Visual inspection - Regularly inspect your antenna for signs of damage or wear. Look for loose or broken connections, damaged elements, or signs of corrosion.

2. Cleaning - Clean your antenna periodically to remove dirt, dust, and other contaminants. Use a soft brush or cloth to gently clean the antenna elements and connections.
3. Weather protection - Protect your antenna from exposure to extreme weather conditions, such as high winds, heavy rain, or snow. Consider installing a weatherproof cover or shelter to protect the antenna.
4. Lubrication - Apply a light coating of a rust inhibitor or other protective lubricant to the antenna elements and connections to prevent corrosion.
5. Grounding - Ensure that your antenna grounding system is properly installed and maintained. Check the ground connections regularly for signs of corrosion or damage.
6. Repair or replacement - If you notice any damage or wear on your antenna, repair or replace it as soon as possible to prevent further damage or interference.

By following these tips for maintaining your antenna, you can ensure optimal performance and reliability during emergency situations and regular ham radio communication.

Modes of Communication

There are several communication modes that can be used for ham radio communication. Here are some of the most common modes:

1. Single Sideband (SSB) - SSB is a voice mode that uses amplitude modulation. It is the most commonly used mode for voice communication on HF frequencies.
2. Frequency Modulation (FM) - FM is a voice mode that uses

frequency modulation. It is commonly used for VHF and UHF communication.

3. Morse code (CW) - Morse code is a digital mode that uses on-off keying to transmit messages. It is commonly used for low-power communication on HF frequencies.
4. Digital modes - There are several digital modes that can be used for ham radio communication, including PSK31, RTTY, and FT8. These modes use digital encoding to transmit messages.
5. Packet radio - Packet radio is a digital mode that allows for the transmission of data messages over radio. It is commonly used for emergency communication and messaging.
6. Voice over Internet Protocol (VoIP) - VoIP allows for voice communication over the internet using ham radio. It is commonly used for long-distance communication.

Each communication mode has its own strengths and weaknesses, and the best choice will depend on the frequency range, bandwidth, and modulation requirements of your communication needs. By understanding the different modes available, you can choose the best mode for your specific communication requirements.

Single Sideband (SSB)

Single Sideband (SSB) is a voice mode that uses amplitude modulation to transmit voice signals. SSB is commonly used for long-range communication on HF frequencies. It is more efficient than other modes, such as AM, as it requires less bandwidth to transmit the same amount of information. This makes it ideal for use during emergency situations where bandwidth may be limited.

SSB signals can be either upper sideband (USB) or lower sideband (LSB), depending on the frequency being used. USB is used for frequencies above 10 MHz, while LSB is used for frequencies below 10 MHz.

To transmit using SSB, you will need a transmitter that is capable of producing an SSB signal, as well as a receiver that is capable of receiving SSB signals. You will also need to use a microphone to input your voice into the transmitter. Once you have established communication with another station, you can adjust the frequency and other settings as necessary to optimize the quality of the transmission.

SSB is an important mode for emergency communication, as it allows for efficient and reliable communication over long distances using minimal bandwidth. By understanding how to use SSB, you can improve your communication capabilities during emergency situations and regular ham radio communication.

Frequency Modulation (FM)

Frequency Modulation (FM) is a voice mode that uses frequency modulation to transmit voice signals. It is commonly used for VHF and UHF communication, such as two way radio communication, amateur radio, and public safety communications.

FM is less susceptible to noise and interference than other modes, such as AM, making it a reliable choice for voice communication. It is also relatively easy to use, as it does not require precise tuning or adjustment like SSB. This makes it a popular choice for handheld radios and mobile communication.

To use FM, you will need a transmitter that is capable of

producing an FM signal, as well as a receiver that is capable of receiving FM signals. You will also need to use a microphone to input your voice into the transmitter. Once you have established communication with another station, you can adjust the frequency and other settings as necessary to optimize the quality of the transmission.

FM is an important mode for emergency communication, as it allows for reliable communication over short to medium distances using VHF and UHF frequencies. By understanding how to use FM, you can improve your communication capabilities during emergency situations and regular ham radio communication.

Morse code (CW)

Morse code, also known as CW (Continuous Wave), is a digital mode that uses on-off keying to transmit messages. It is a simple and reliable mode that has been used for communication since the early days of radio.

Morse code uses a combination of dots and dashes to represent letters, numbers, and punctuation. Each character has a unique sequence of dots and dashes that can be easily memorized and transmitted using a simple keyer or paddle.

Morse code is commonly used for low-power communication on HF frequencies, as it is more efficient than other modes for low-power transmission. It is also a useful mode for emergency communication, as it can be used to transmit messages in low-bandwidth and noisy environments.

To use Morse code, you will need a transmitter that is capable of producing a CW signal, as well as a receiver that is capable of receiving CW signals. You will also need a keyer or paddle to input your message into the transmitter. Once you have established communication with another station, you can adjust the frequency and other settings as necessary to optimize the quality of the transmission.

Morse code is an important mode for emergency communication, as it allows for reliable communication in low-bandwidth and noisy environments. By understanding how to use Morse code, you can improve your communication capabilities during emergency situations and regular ham radio communication.

Digital modes

Digital modes are a group of communication modes that use digital encoding to transmit messages over radio. These modes are efficient and reliable, as they are less susceptible to noise and interference than voice modes.

Some of the most common digital modes used in ham radio communication include:

1. PSK31 - PSK31 is a digital mode that uses phase-shift keying to transmit data at a speed of 31 baud. It is commonly used for low-power communication on HF frequencies.
2. RTTY - RTTY (Radio Teletype) is a digital mode that uses frequency-shift keying to transmit data. It is commonly used for high-speed data communication on HF frequencies.
3. FT8 - FT8 is a digital mode that uses frequency-shift keying to transmit data at a high rate of speed. It is commonly used for long-distance communication on HF

frequencies.

To use digital modes, you will need a computer with a sound card and a specialized software program that can decode and encode the digital signals. You will also need a radio that is capable of transmitting and receiving digital signals, as well as a specialized interface that connects your computer to the radio.

Digital modes are an important tool for emergency communication, as they allow for efficient and reliable communication over low-bandwidth and noisy channels. By understanding how to use digital modes, you can improve your communication capabilities during emergency situations and regular ham radio communication.

Packet radio

Packet radio is a digital mode that allows for the transmission of data messages over radio. It is a reliable and efficient mode that is commonly used for emergency communication and messaging.

Packet radio uses a system of packet switching, similar to the internet, to send and receive data. This allows for more efficient use of the radio frequency and enables multiple transmissions to take place at the same time.

To use packet radio, you will need a specialized packet radio modem that can encode and decode digital signals. You will also need a computer and a specialized software program that can communicate with the modem and enable you to compose and send messages. Once you have established communication with another station, you can send and receive data messages in real-time.

Packet radio is an important mode for emergency communication, as it allows for efficient and reliable communication of critical information, such as weather reports, emergency messages, and other important data. By understanding how to use packet radio, you can improve your communication capabilities during emergency situations and regular ham radio communication.

Voice over Internet Protocol (VoIP)

Voice over Internet Protocol (VoIP) is a communication mode that allows for voice communication over the internet. It is a digital mode that uses the internet to transmit voice signals, rather than traditional radio frequencies.

VoIP is commonly used for long-distance communication, as it is more cost-effective than traditional phone lines. It is also a useful mode for emergency communication, as it allows for communication over long distances without the need for specialized radio equipment.

To use VoIP, you will need a computer or a smartphone with an internet connection and a specialized software program, such as Skype or Zoom, that can transmit and receive voice signals. Once you have established communication with another station, you can communicate in real-time using voice communication.

VoIP is an important mode for emergency communication, as it allows for communication over long distances without the need for specialized radio equipment. However, it is important to note that VoIP may be subject to internet disruptions or outages, which can impact the reliability of communication during emergency situations. Therefore, it is recommended to have multiple communication modes, including traditional radio communication, as a backup in case of internet disruptions.

Low Cost Mobile Amateur Radio Station

Equipment Selection

For a low-cost amateur radio station for communicating statewide, you may consider the following equipment:

1. Transceiver: The Baofeng UV-5R is a low-cost, versatile handheld transceiver that can transmit on the VHF and UHF bands. It is a popular option for beginners and experienced hams alike, and can be found for under $30.
2. Antenna: The Nagoya NA-771 is a popular upgrade antenna for the Baofeng UV-5R. It is a flexible, lightweight, and relatively inexpensive option that can improve your signal transmission and reception.
3. Power Supply: The Pyramid PS9KX is a small, affordable, and reliable power supply that can provide up to 5 amps of power. It is suitable for powering your transceiver and other accessories.
4. Coaxial Cable: You will need a coaxial cable to connect your antenna to your transceiver. The MPD Digital LMR-400 is a high-quality, low-loss coaxial cable that can provide good signal transmission and reception.
5. SWR Meter: The Astatic PDC1 SWR Meter is an inexpensive and easy-to-use meter that can help you tune your antenna for optimal performance.

When it comes to brand selection, some reputable brands in the amateur radio equipment industry include Yaesu, Icom, Kenwood, and Alinco. However, for a low-cost setup, you may find that the Baofeng and Nagoya brands offer good value for the price. As always, be sure to do your research and read reviews before making any purchases.

Step-By-Step Setup

Here are step-by-step instructions for setting up a low-cost amateur radio station for statewide communication:

1. Choose a reliable, low-cost mobile or base transceiver: A few popular and affordable options include the Yaesu FT-857D, Icom IC-7100, and Kenwood TM-V71A.
2. Purchase a suitable antenna: For statewide communication, a high-gain antenna is recommended. Some popular options include the Comet GP-9, Diamond X50A, and Nagoya UT-308UV.
3. Install the transceiver and antenna: Follow the manufacturer's instructions for installing the transceiver and antenna. Ensure that the antenna is securely grounded to prevent any potential electrical hazards.
4. Purchase a power supply: For a mobile transceiver, a 12-volt power supply is recommended, while a 110-volt power supply is needed for a base transceiver.
5. Set up a power source: Connect the power supply to a reliable power source, such as a wall outlet or a car battery. Ensure that the power source can handle the voltage requirements of the transceiver.
6. Test your equipment: Before attempting to make any con-

tacts, test your equipment to ensure that it is functioning correctly. You can test the antenna by checking its SWR (Standing Wave Ratio) with an SWR meter.

7. Tune in to the statewide repeater: Find the statewide repeater's frequency by checking online databases or contacting local ham radio clubs. Tune your transceiver to the frequency and ensure that you are using the correct CTCSS (Continuous Tone-Coded Squelch System) tone.
8. Listen for incoming signals: Once you have tuned in to the repeater frequency, listen for any incoming signals. Make sure that your volume is set to an appropriate level.
9. Make contact: When you hear a signal, respond with your callsign and location. If the other operator responds, you can then engage in a conversation.
10. Practice good operating practices: Ensure that you are following all FCC regulations and operating practices, such as avoiding the use of offensive language and not transmitting excessively.

With these steps, you should be able to set up a low-cost amateur radio station for statewide communication. Remember to always practice good operating practices and ensure that you are operating within the guidelines set by the FCC.

Medium Cost Amateur Radio Base Station

Equipment Selection

A medium-cost amateur radio station for state-wide communication using a base station could consist of the following equipment:

1. Transceiver: Yaesu FT-857D or Kenwood TS-480SAT ($1000-$1200)
2. Antenna: Diamond CP-6AR multi-band vertical antenna or Comet GP-9N dual-band vertical antenna ($250-$350)
3. Coaxial cable: Belden RG-8X or LMR-400 ($75-$100)
4. Power supply: Astron RS-35M or Samlex SEC-1235M ($150-$200)
5. Tuner: LDG AT-100ProII or MFJ-939Y ($200-$250)

Step-By-Step Setup

Here are step-by-step instructions for setting up the medium-cost amateur radio station:

1. Choose a location for your base station: A good location for your base station would be a place where you have space to set up your antenna and enough room to operate your equipment comfortably.
2. Install your antenna: Mount your antenna to a suitable mast or tower, and run your coaxial cable from the antenna to your base station.
3. Set up your transceiver: Connect your transceiver to your power supply, antenna, and microphone. Power on the transceiver and set it to the desired frequency and mode.
4. Install your tuner: If using a multi-band antenna, you may need to install a tuner to match the impedance of the antenna to the transceiver. Connect your tuner between your transceiver and the coaxial cable running to the antenna.
5. Test your equipment: Tune your transceiver to a known frequency and check the SWR reading on your tuner to

ensure it is below 2:1. Test your microphone, headphones, and key to make sure they are functioning properly.

6. Set up communication frequencies: Determine the frequencies used by other ham radio operators in your area and program them into your transceiver. You can also set up memory channels for easy access to frequently used frequencies.
7. Test your communication: Test your communication by calling out on the frequencies you programmed into your transceiver. Listen for responses from other ham radio operators and adjust your equipment as necessary.

With this medium-cost setup, you should be able to communicate with other ham radio operators throughout your state and beyond.

6

Operating During Emergencies

The Union City, Oklahoma F4 tornado on May 24, 1973

Signal strength and propagation are important factors to consider when setting up your ham radio station for emergency communication. The strength of the signal that you transmit and receive can impact the clarity and reliability of your communication, especially during emergency situations.

Signal strength is affected by a number of factors, including the power output of your radio transmitter, the quality of your antenna, the frequency you are using, and the environment around you. The higher the power output of your transmitter and the better the quality of your antenna, the stronger your signal will be. The frequency you are using also plays a role, as some frequencies are better suited for long-distance communication than others.

Propagation refers to the way in which radio waves travel through the environment. Radio waves can be affected by various factors, including terrain, atmospheric conditions, and the time of day. For example, radio waves at lower frequencies can travel further over the ground, while radio waves at higher frequencies can travel further through the atmosphere.

To optimize signal strength and propagation during emergency communication, it is important to consider factors such as antenna height, location, and grounding, as well as the frequency you are using and the time of day. It is also important to understand how to adjust your transmitter settings to optimize signal strength and clarity.

By understanding how signal strength and propagation work, and by taking steps to optimize your station and your communication techniques, you can improve the reliability and effectiveness of your emergency communication during critical situations.

Maximizing signal strength

Maximizing signal strength is an important consideration when setting up your ham radio station for emergency communication. A strong signal will allow you to communicate more effectively over long distances, improving the reliability of your communication during critical situations.

To maximize signal strength, there are several factors to consider:

1. Antenna: A high-quality antenna is essential for maximizing signal strength. Choose an antenna that is appropriate for the frequency you will be using, and ensure that it is properly installed and grounded.
2. Height: The higher your antenna is, the better your signal will be. Consider installing your antenna as high as possible, such as on a tall mast or on a rooftop.
3. Power: Increasing the power output of your transmitter can improve your signal strength, but be aware of regulatory limits and potential interference issues.
4. Frequency: Some frequencies are better suited for long-distance communication than others. Research the best frequency to use for your specific situation.
5. Line-of-sight: The fewer obstacles between your antenna and the receiving station, the stronger your signal will be. Consider the location and terrain around your station and try to find a clear line-of-sight to the receiving station.
6. Radio settings: Adjusting your radio settings, such as the squelch level and the modulation, can also help to improve signal strength.

By considering these factors and taking steps to optimize your station, you can maximize your signal strength and improve the reliability of your communication during emergency situations.

Understanding propagation

Propagation is the way in which radio waves travel through the environment. Understanding propagation is important for optimizing your ham radio station for emergency communication, as it can affect the range and reliability of your signal.

Radio waves can be affected by a variety of factors, including terrain, atmospheric conditions, and the frequency you are using. The main modes of propagation include:

1. Ground wave propagation: This mode of propagation is used for short-range communication and occurs when radio waves travel along the surface of the earth.
2. Sky wave propagation: This mode of propagation is used for long-range communication and occurs when radio waves are reflected off the ionosphere and back down to earth. This mode is affected by atmospheric conditions, such as the density and height of the ionosphere, as well as the frequency of the radio wave.
3. Line-of-sight propagation: This mode of propagation occurs when radio waves travel in a straight line between two stations that are in sight of each other. This mode is limited by the curvature of the earth and obstacles, such as buildings and mountains.

Understanding the different modes of propagation and how

they are affected by different factors can help you to optimize your station for emergency communication. For example, if you need to communicate over long distances, you may need to use a higher frequency to take advantage of sky wave propagation. Alternatively, if you are communicating over short distances, you may need to consider line-of-sight propagation and adjust your antenna height and location accordingly.

By understanding propagation, you can optimize your ham radio station for emergency communication and improve the reliability and effectiveness of your communication during critical situations.

Communication protocols

Communication protocols are a set of rules that govern the exchange of information between two or more stations. In emergency communication, protocols are important for ensuring clear and effective communication, especially in high-stress situations where time is of the essence.

There are several key communication protocols that are commonly used in emergency communication, including:

1. Standard Operating Procedures (SOPs). These are standard protocols for operating and communicating during emergency situations. SOPs are designed to ensure that communication is clear, concise, and effective, and that all parties involved understand their roles and responsibilities.
2. Phonetic Alphabet: The phonetic alphabet is a standardized set of words used to represent each letter of the

alphabet. This is important for spelling out words or names over the radio, especially when communication is affected by poor signal quality or interference.

3. Prowords: Prowords are short words or phrases used to convey specific information, such as "Roger" to indicate that a message has been received, or "Over" to indicate that it is the other station's turn to speak.
4. Q Codes: Q Codes are standardized three-letter codes used to communicate specific information, such as "QSL" to confirm that a message has been received and understood.
5. Emergency Communication Plans: These are pre-established plans for communication during emergency situations. They may include protocols for calling for help, coordinating with other stations or agencies, and relaying critical information.

By following these communication protocols, you can ensure that your communication is clear, effective, and efficient during emergency situations. It is important to practice and review these protocols regularly to ensure that you are prepared to communicate effectively when it matters most.

Phonetic Alphabet

The Phonetic Alphabet, also known as the International Radiotelephony Spelling Alphabet, is a standardized set of words used to represent each letter of the alphabet. This is important for spelling out words or names over the radio, especially when communication is affected by poor signal quality or interference.

The Phonetic Alphabet is commonly used in emergency

communication and aviation communication, as well as by military and law enforcement personnel. Each letter of the alphabet is represented by a unique word, as follows:

A - Alpha
B - Bravo
C - Charlie
D - Delta
E - Echo
F - Foxtrot
G - Golf
H - Hotel
I - India
J - Juliet
K - Kilo
L - Lima
M - Mike
N - November
O - Oscar
P - Papa
Q - Quebec
R - Romeo
S - Sierra
T - Tango
U - Uniform
V - Victor
W - Whiskey
X - X-ray
Y - Yankee
Z - Zulu

By using the Phonetic Alphabet, you can ensure that your communication is clear and accurate, even when signal quality is poor or there is interference. It is important to practice and review the Phonetic Alphabet regularly to ensure that you are prepared to communicate effectively during emergency situations.

Prowords

Prowords, also known as procedural words or prosigns, are short words or phrases used to convey specific information during communication. They are commonly used in emergency communication and other situations where clear and concise communication is essential.

Here are some examples of common prowords:

1. Roger: This word is used to indicate that a message has been received and understood.
2. Over: This word is used to indicate that it is the other station's turn to speak.
3. Out: This word is used to indicate the end of a transmission or conversation.
4. Wilco: This word is used to indicate that a message has been received, understood, and will be complied with.
5. Say Again: This phrase is used to request the other station to repeat a message or information.
6. Break: This word is used to indicate a pause in transmission, typically followed by important information.

By using prowords, you can ensure that your communication is clear, concise, and efficient during emergency situations. It is important to practice and review these prowords regularly to ensure that you are prepared to communicate effectively when it matters most.

Q Codes

Q Codes are a standardized set of three-letter codes used to communicate specific information during radio communication. They were originally developed for maritime communication and have since been adopted by other radio communication services, including ham radio and aviation communication.

Q Codes cover a wide range of topics and situations, including weather conditions, signal quality, equipment issues, and more. Here are some examples of common Q Codes:

1. QRL: Are you busy?
2. QRM: Are you being interfered with?
3. QRN: Are you experiencing static or noise?
4. QRP: Are you using low power?
5. QRT: Shall I stop sending?
6. QSL: Can you confirm receipt?
7. QSY: Shall I change frequency?
8. QTH: What is your location?

By using Q Codes, you can communicate information quickly and efficiently, especially in emergency situations where time is critical. It is important to familiarize yourself with the common Q Codes and to practice using them regularly to ensure that you are prepared to communicate effectively during emergency situations.

Emergency Communication Plan

An Emergency Communication Plan is a pre-established plan for communication during emergency situations. It is designed to ensure that communication is clear, concise, and effective, and that all parties involved understand their roles and responsibilities.

Emergency Communication Plans are essential for emergency responders, public safety agencies, and individuals who may need to communicate during emergency situations. Here are some key components of an Emergency Communication Plan:

1. Contact Information: This includes contact information for all members of the communication team, as well as for emergency responders and other relevant agencies.
2. Communication Protocols: This includes the communication protocols that will be used during emergency situations, such as standard operating procedures, phonetic alphabet, prowords, and Q codes.
3. Radio Frequencies: This includes the radio frequencies that will be used during emergency situations, as well as backup frequencies in case of interference or other issues.
4. Equipment and Power Sources: This includes a list of equipment that will be used during emergency situations, as well as backup power sources in case of power outages.
5. Roles and Responsibilities: This includes a clear definition of the roles and responsibilities of each member of the communication team, as well as any emergency respon-

ders or other agencies involved in the communication plan.

By creating an Emergency Communication Plan, you can ensure that you are prepared to communicate effectively during emergency situations. It is important to review and practice the plan regularly to ensure that all parties involved understand their roles and responsibilities and are prepared to communicate effectively when it matters most.

Net procedures

Net procedures are a set of guidelines and protocols used during net operations in emergency communication situations. A net is a structured communication system used to coordinate and manage communication during an emergency, and can involve multiple stations or participants.

Here are some key elements of net procedures:

1. Net Control Station (NCS): The NCS is responsible for coordinating the communication and directing the flow of information during the net. They are responsible for starting and ending the net, and for managing the communication between stations.
2. Check-In Procedures: Each station should check in with the NCS at the beginning of the net and provide their call sign, location, and any other relevant information.
3. Traffic Handling: The NCS is responsible for managing the flow of information and directing the traffic to the appropriate station.

4. Emergency Priority Traffic: Emergency priority traffic takes precedence over all other traffic and should be identified as such during the net. The NCS is responsible for prioritizing and directing emergency traffic.
5. Formal Traffic: Formal traffic involves important information that needs to be recorded and tracked. The NCS is responsible for managing formal traffic and directing it to the appropriate station.
6. Closing Procedures: The NCS is responsible for ending the net and ensuring that all stations have checked out and are accounted for.

By following these net procedures, communication during emergency situations can be managed more efficiently and effectively, allowing for a more coordinated response to the emergency.

Emergency Traffic Handling

Emergency traffic handling is the process of managing and prioritizing communication related to emergency situations. During emergency situations, it is critical to ensure that emergency traffic is handled quickly and efficiently to ensure that life-saving information is received and acted upon in a timely manner.

Here are some key elements of emergency traffic handling:

1. Prioritization: Emergency traffic should be given the highest priority and handled before any other traffic.
2. Identification: Emergency traffic should be clearly identi-

fied as such to ensure that it is recognized and prioritized appropriately.

3. Direct Routing: Emergency traffic should be routed directly to the appropriate station or authority as quickly as possible to ensure that it is received and acted upon in a timely manner.
4. Relay Stations: In situations where direct communication is not possible, relay stations can be used to forward emergency traffic to the appropriate authority.
5. Confirmation of Receipt: The sender of emergency traffic should receive confirmation that the message has been received and acted upon.

By following these emergency traffic handling procedures, emergency communication can be managed more efficiently and effectively, allowing for a more coordinated response to the emergency. It is important to practice and review these procedures regularly to ensure that you are prepared to communicate effectively during emergency situations.

Communications Mode

Using different communication modes is an important aspect of emergency communication. Different communication modes allow you to communicate with other stations or authorities that may not be reachable through other means.

Here are some common communication modes used in emergency communication:

1. Voice Modes: Voice modes such as Single Sideband (SSB)

and Frequency Modulation (FM) allow for clear and efficient voice communication over long distances.
2. Morse Code: Morse code is a time-tested mode that is still used by some ham radio operators. It is particularly useful when voice communication is difficult or not possible.
3. Digital Modes: Digital modes such as packet radio and Winlink allow for the transmission of data and text messages. These modes are particularly useful for sending important information such as maps, photos, and status updates.
4. Satellite Communication: Satellite communication can be used to communicate with remote locations or areas where other communication modes are not available.

By using different communication modes, you can increase your communication capabilities and improve your chances of communicating with other stations or authorities during emergency situations. It is important to practice using these different modes and to ensure that your equipment is configured to work with each mode.

Morse Code

Morse code is a system of dots and dashes that can be transmitted over the airwaves using a telegraph key or other input device. It is named after its inventor, Samuel Morse, and was originally used for long-distance communication over telegraph lines.

Today, Morse code is still used by some ham radio operators, particularly in emergency communication situations where other modes of communication may not be available or reliable.

Morse code is particularly useful because it can be transmitted using low power and is resistant to interference.

Here are some key elements of Morse code:

1. Letters and Numbers: Morse code consists of a series of dots and dashes that represent letters and numbers. For example, the letter A is represented by a dot followed by a dash, while the number 1 is represented by a single dot.
2. Speed: Morse code can be sent at different speeds, ranging from slow (5 words per minute) to fast (40 words per minute). It is important to practice and become proficient at different speeds in order to communicate effectively using Morse code.
3. Prosigns: Prosigns are special codes used in Morse code to convey important information such as the end of a message or the request for a repeat.
4. Farnsworth Timing: Farnsworth timing is a method of learning Morse code that involves learning the individual characters at a slow speed but with a normal spacing between characters.

By learning Morse code and practicing its use, ham radio operators can improve their communication capabilities and be better prepared to communicate in emergency situations where other modes of communication may not be available.

Morse Code Alphabet

Here is the Morse code alphabet spelled out using the words "dot" and "dash":

- A: dot-dash
- B: dash-dot-dot-dot
- C: dash-dot-dash-dot
- D: dash-dot-dot
- E: dot
- F: dot-dot-dash-dot
- G: dash-dash-dot
- H: dot-dot-dot-dot
- I: dot-dot
- J: dot-dash-dash-dash
- K: dash-dot-dash
- L: dot-dash-dot-dot
- M: dash-dash
- N: dash-dot
- O: dash-dash-dash
- P: dot-dash-dash-dot
- Q: dash-dash-dot-dash
- R: dot-dash-dot
- S: dot-dot-dot
- T: dash
- U: dot-dot-dash
- V: dot-dot-dot-dash
- W: dot-dash-dash
- X: dash-dot-dot-dash
- Y: dash-dot-dash-dash
- Z: dash-dash-dot-dot

By learning and practicing the Morse code alphabet, ham radio operators can improve their communication capabilities and be better prepared to communicate in emergency situations where other modes of communication may not be available.

7

Working with Others

Coordinating with First Responders

Coordination with Other Operators

Ham radio operators can work with emergency management officials to improve communication and response during emergency situations. This section will cover key aspects of collaboration between ham radio operators and emergency management officials, including establishing relationships, understanding emergency management structures and protocols, developing communication plans, participating in emergency exercises and drills, and incorporating feedback.

1. Establishing Relationships:

- Building relationships with emergency management officials is an important aspect of collaboration.
- Introducing yourself to local officials, attending community meetings and events, and participating in emergency exercises and drills are all ways to establish relationships.
- Building trust and understanding the role of emergency management officials can help facilitate effective communication during emergency situations.

1. Understanding Emergency Management Structures and Protocols.

- Understanding the structure and protocols of emergency management organizations is crucial for effective collaboration.
- Familiarize yourself with the emergency management

structure in your area and learn about the different roles and responsibilities of officials.
- Understand the protocols for declaring emergencies, activating emergency response plans, and coordinating resources.

1. Developing Communication Plans:

- Developing communication plans with emergency management officials can help ensure that communication efforts are well-coordinated and effective.
- Collaborate with emergency management officials to identify communication needs and develop plans for how ham radio operators can support communication efforts.
- Communication plans should include identifying communication channels, establishing protocols for relaying messages, and coordinating frequencies and modes of communication.

1. Participating in Emergency Exercises and Drills:

- Participating in emergency exercises and drills is a great way to gain experience working with emergency management officials and improve your communication skills.
- During exercises and drills, ham radio operators can practice communication protocols and work with emergency management officials to identify areas for improvement.
- Exercising your communication plan can help identify potential issues and improve communication effectiveness.

1. Incorporating Feedback:

- Incorporating feedback from emergency management officials is essential for ongoing improvement of communication plans and strategies.
- Seek feedback from emergency management officials following emergency situations, exercises, and drills.
- Incorporate feedback into your communication plan and make necessary adjustments to improve effectiveness.

Collaboration between ham radio operators and emergency management officials is crucial for effective emergency communication and response. By establishing relationships, understanding emergency management structures and protocols, developing communication plans, participating in emergency exercises and drills, and incorporating feedback, ham radio operators can play a vital role in emergency response efforts.

Collaboration with First Responders

Ham radio operators play an important role in emergency communication by supporting first responders such as police officers, firefighters, and medical personnel. Collaborating with first responders is critical in emergency communication, and ham radio operators can achieve this through various means:

1. Establishing relationships: Building trust and establishing a relationship with first responders is important. By introducing yourself to local first responders and participating in emergency response training exercises, you can establish a good working relationship with them, which can be essential in emergency situations.

2. Understanding communication needs: Ham radio operators need to understand the communication protocols used by first responders and their equipment capabilities. By doing so, they can tailor their communication efforts to meet their needs. Ham radio operators can also work with first responders to develop communication plans that can be used in emergency situations.
3. Working together during emergency situations: Collaboration between ham radio operators and first responders is put to the ultimate test during emergency situations. Ham radio operators can work with first responders to relay critical information, coordinate search and rescue efforts, and provide communication support.
4. Incorporating feedback: First responders can provide valuable feedback to ham radio operators regarding the effectiveness of their communication efforts. By asking for feedback and incorporating it into your communication plan, ham radio operators can improve their communication efforts and ensure that they are meeting the needs of first responders.
5. Participating in emergency response exercises: Ham radio operators can gain valuable experience working with first responders and improve their communication skills by participating in emergency response exercises and drills.

In conclusion, collaboration with first responders is essential in emergency communication. By establishing relationships, understanding communication needs, working together during emergency situations, incorporating feedback, and participating in emergency response exercises, ham radio operators can play a critical role in supporting first responders and ensuring effective communication in emergency situations.

Partnership with Emergency Management Officials

Emergency management officials play a key role in coordinating emergency response efforts and ensuring that resources are allocated effectively during times of crisis. Ham radio operators can work in partnership with emergency management officials to ensure that communication efforts are well coordinated and that critical information is relayed quickly and accurately. This section will cover topics such as understanding the role of emergency management officials, communicating with emergency management officials, coordinating emergency response efforts, and best practices for working with emergency management officials.

Partnership with emergency management officials is critical for effective emergency communication. This partnership can be achieved by building relationships, understanding communication needs, working together during emergency situations, and incorporating feedback.

1. Building relationships with emergency management officials is key to establishing trust and effective collaboration. Introducing yourself, participating in emergency response training exercises, and establishing a good working relationship are important steps in building this relationship.
2. Understanding the communication needs of emergency management officials is also essential. By understanding the communication protocols used by emergency management officials and their equipment capabilities, ham radio operators can tailor their communication efforts to

meet their needs. Developing communication plans with emergency management officials can also help ensure that everyone is on the same page in emergency situations.

3. Working together during emergency situations is the ultimate test of partnership. Ham radio operators can support emergency management officials by relaying critical information, coordinating emergency response efforts, and providing communication support. Participation in emergency response exercises and drills can help ham radio operators gain valuable experience working with emergency management officials and improve their communication skills.
4. Feedback from emergency management officials can help ham radio operators improve their communication efforts. By asking for feedback and incorporating it into their communication plan, ham radio operators can ensure that they are meeting the needs of emergency management officials. Regular debriefings after emergency situations can also provide valuable insights into what worked well and what could be improved.

In summary, partnership with emergency management officials is critical in emergency communication. Building relationships, understanding communication needs, working together during emergency situations, and incorporating feedback are all essential elements of this partnership. By working together, ham radio operators and emergency management officials can ensure that communication is effective and coordinated in emergency situations.

Mutual Aid Agreements

Mutual aid agreements are formal agreements between organizations that provide assistance to each other during times of crisis. Ham radio operators can establish mutual aid agreements with other organizations to help ensure effective communication during emergency situations.

Collaborating with other organizations can help to expand the resources available during emergency situations. Through mutual aid agreements, ham radio operators can partner with other organizations to share resources, equipment, and knowledge. This can help to improve emergency response efforts and ensure that critical information is relayed quickly and accurately.

Developing and implementing mutual aid agreements involves several key steps:

1. Identify potential partners: Ham radio operators should identify potential partner organizations that they can work with during emergency situations. These may include other ham radio groups, emergency management agencies, and local government agencies.
2. Establish mutual aid agreements: Once potential partners have been identified, mutual aid agreements can be established. These agreements should outline the roles and responsibilities of each party and the resources that will be shared during emergency situations.
3. Communicate and train: Effective communication is essential to successful collaboration. Ham radio operators should communicate regularly with their partner orga-

nizations to ensure that everyone is informed of updates and changes. Training exercises can also be conducted to ensure that everyone is prepared to work together during emergency situations.

4. Regular review and update: Mutual aid agreements should be regularly reviewed and updated to ensure that they remain effective and relevant. Changes in technology, equipment, and personnel can all impact the effectiveness of mutual aid agreements, and it is important to ensure that these agreements are kept up to date.

Best practices for mutual aid agreements include:

1. Establish clear roles and responsibilities: All parties should have a clear understanding of their roles and responsibilities during emergency situations. This can help to ensure that everyone is working towards the same goal and that critical tasks are not overlooked.
2. Develop effective communication channels: Communication is essential to effective collaboration. Ham radio operators should work with their partner organizations to establish effective communication channels and protocols.
3. Identify key resources: Ham radio operators should identify key resources that can be shared with their partner organizations. This may include equipment, personnel, or expertise.
4. Plan for contingencies: Contingency planning is essential to effective emergency response. Ham radio operators should work with their partner organizations to plan for contingencies and unexpected events.

Overall, mutual aid agreements can help to improve emergency response efforts and ensure that critical information is relayed quickly and accurately. By establishing relationships with partner organizations, developing effective communication channels, and sharing resources, ham radio operators can play a vital role in emergency response and communication.

Coordinating frequencies

Coordinating frequencies is a crucial aspect of effective emergency communication among ham radio operators. Here are some key points to keep in mind:

1. Understanding frequency coordination: Coordinating frequencies involves identifying the frequencies that will be used by different operators in a given area to avoid interference and ensure that communication efforts are well-organized. This can be done through various means such as a local frequency coordinator or communication plans developed by emergency management officials.
2. Identifying available frequencies: Ham radio operators must identify the available frequencies for use in their local area. This may involve working with local radio clubs or contacting other ham radio operators to determine which frequencies are available for use.
3. Developing communication plans: Once frequencies have been identified, ham radio operators must develop communication plans to ensure that communication efforts are well-coordinated. This may involve developing protocols for different frequencies and modes, and identifying net control operators who can manage communication

efforts.

4. Maintaining flexibility: In emergency situations, frequencies may become congested or unavailable, so it is important for ham radio operators to be flexible and willing to adjust their communication plans as needed. This may involve switching frequencies or modes or adapting to changing circumstances.
5. Coordinating with other emergency responders: Ham radio operators must also coordinate their frequency use with other emergency responders such as police, fire, and medical personnel. This may involve working with local emergency management officials to ensure that communication plans are well-coordinated and that communication efforts are not interfering with those of other responders.

By effectively coordinating frequencies, ham radio operators can ensure that communication efforts are well-organized and that emergency messages are relayed quickly and accurately.

Situational awareness

Situational awareness is an important aspect of emergency communication. By providing situational awareness, ham radio operators can help emergency responders and officials understand the scope and severity of a crisis, which can inform their response efforts. Here are some ways that ham radio operators can provide situational awareness:

1. Relay information: Ham radio operators can relay information about the situation on the ground, such as the

extent of damage, the number of casualties, and the status of rescue and recovery efforts.

2. Monitor news and social media: Ham radio operators can monitor news and social media for information about the emergency situation and relay this information to emergency responders and officials.
3. Provide weather updates: Ham radio operators can provide real-time weather updates, including information about severe weather events, such as tornadoes, hurricanes, and floods, that can impact emergency response efforts.
4. Monitor traffic and transportation: Ham radio operators can monitor traffic and transportation systems, such as highways, airports, and railroads, to provide updates on road and bridge closures, transportation delays, and other factors that can impact emergency response efforts.
5. Share maps and other data: Ham radio operators can share maps and other data, such as satellite imagery, that can help emergency responders and officials understand the extent of the crisis and identify areas that need assistance.

Overall, providing situational awareness is an important role that ham radio operators can play in emergency communication. By relaying information, monitoring news and social media, providing weather updates, monitoring traffic and transportation, and sharing maps and other data, ham radio operators can help emergency responders and officials make informed decisions and respond effectively to crises.

8

Advanced Techniques

Communications Satellite

Advanced techniques in ham radio communication can enhance the efficiency and effectiveness of communication during emergency situations. These techniques go beyond the basic operating procedures and equipment considerations discussed in earlier chapters, and can help ham radio operators to communicate more effectively with other operators, first responders, and emergency officials. Some advanced techniques that ham radio operators can use to improve their communication capabilities include digital modes such as PSK31, using software-defined radios (SDRs), and employing satellite communication.

Digital modes such as PSK31 are one way that ham radio operators can expand their communication capabilities. PSK31 is a popular digital mode that uses phase-shift keying (PSK) to send data over the airwaves. Compared to traditional modes like voice and Morse code, digital modes can be more efficient and reliable, allowing for more data to be transmitted in less time. Additionally, digital modes like PSK31 can be decoded using computer software, making them more accessible to new ham radio operators.

Software-defined radios (SDRs) are another advanced technique that can enhance communication capabilities. SDRs allow operators to tune and adjust their radios using software, rather than using physical knobs and switches. This allows for greater flexibility and precision when it comes to tuning in to specific frequencies and modes. Additionally, SDRs can be programmed to perform specific tasks automatically, freeing up operators to focus on other aspects of emergency communication.

Satellite communication is a third advanced technique that can be used by ham radio operators to enhance their communication capabilities. By using amateur radio satellites, operators can communicate with other operators around the world, as well as with first responders and emergency officials in remote areas. While satellite communication requires specialized equipment and technical knowledge, it can be an invaluable tool for communicating during emergency situations where traditional communication infrastructure may be compromised.

Overall, the use of advanced techniques in ham radio communication can greatly enhance the ability of operators to communicate during emergency situations. By employing digital modes like PSK31, using software-defined radios, and leveraging satellite communication, ham radio operators can expand their communication capabilities and help ensure that critical information is relayed quickly and accurately during times of crisis.

Digital modes

Digital modes are becoming increasingly popular among ham radio operators due to their efficiency, reliability, and ability to transmit data over long distances. This section will cover some advanced techniques for using digital modes, including:

1. Operating multiple modes simultaneously: One advantage of digital modes is the ability to operate multiple modes simultaneously on the same band. By using software such as Fldigi, ham radio operators can transmit and receive messages using different modes at the same time. This can help maximize the use of available bandwidth and improve communication efficiency.
2. Developing macros and templates: Digital modes often involve sending repetitive messages or exchanging standardized information. Developing macros and templates can help automate this process, saving time and improving accuracy. Ham radio operators can use software such as

Fldigi to create macros and templates for commonly used messages and information.

3. Using sound card interfaces: Sound card interfaces allow ham radio operators to connect their radio to their computer and transmit digital modes directly from the computer. This can help simplify the setup process and improve the quality of the transmission.
4. Using digital modes for emergency messaging: Digital modes can be used to transmit emergency messages quickly and efficiently. Ham radio operators can use software such as Winlink to send email-style messages over the airwaves, even in areas without internet or phone service.
5. Exploring new digital modes: The world of digital modes is constantly evolving, with new modes being developed and released regularly. Ham radio operators can stay up-to-date on the latest digital modes by reading ham radio publications and forums, attending conferences and events, and experimenting with different software and equipment.

By using these advanced techniques for digital modes, ham radio operators can improve their communication efficiency and effectiveness, particularly in emergency situations.

PSK31

PSK31 is a popular digital mode used by ham radio operators for communication. It is a narrow-bandwidth digital mode that allows for high-speed communication and efficient use of the radio spectrum. Here are some advanced techniques for using

PSK31:

1. Keyboard-to-Keyboard Chatting: PSK31 allows for keyboard-to-keyboard chatting, which means that operators can communicate with each other in real-time using a computer keyboard. This feature can be particularly useful during contests or emergency situations when quick and efficient communication is necessary.
2. Macros: PSK31 software often allows for the creation of macros, which are pre-defined messages that can be sent with a single click of a button. Macros can be used to quickly send commonly used messages, such as "QSL" (acknowledgment of receipt) or "73" (best wishes).
3. Call Sign Lookup: Many PSK31 software programs have a call sign lookup feature, which allows operators to quickly find information about the other station they are communicating with. This can include the other operator's name, location, and other details.
4. Digital Signal Processing: PSK31 relies on digital signal processing (DSP) technology to decode signals. Advanced PSK31 software can use DSP technology to filter out unwanted noise and interference, making communication more efficient and reliable.
5. QSO Logging: PSK31 software often includes a logging feature that can automatically log QSOs (contacts) with other stations. This can be useful for keeping track of contacts during contests or for record-keeping purposes.

Overall, PSK31 is a versatile digital mode that can be used for both casual chatting and more advanced communication needs. By utilizing advanced techniques such as keyboard-to-keyboard chatting, macros, call sign lookup, DSP, and QSO logging, operators can maximize the benefits of PSK31 for their communication needs.

Winlink

Winlink is an advanced digital mode that allows ham radio operators to send and receive email messages over radio waves, even when there is no internet or phone service available. It is a popular method of communication for emergency response teams, as it enables them to send and receive critical information, including weather reports, damage assessments, and rescue plans, in real-time.

One of the benefits of Winlink is its ability to work on low power, making it an ideal solution in emergency situations where power may be limited. It is also a flexible mode that can be used with a variety of radio equipment and can transmit messages over long distances.

Winlink operates using a client-server architecture, with a central server that acts as a gateway to the internet. Ham radio operators can connect to this server using their radio equipment and send and receive email messages. The system uses various modes of transmission, including Pactor, a high speed modem technology that can operate even in poor signal conditions.

To use Winlink, ham radio operators must first register for an account on the system's website. They can then download and install software on their computer or mobile device to connect to the Winlink server. Once connected, they can send and receive email messages, as well as access various resources, such as weather reports and emergency bulletins.

In addition to email, Winlink can also be used to send and receive other types of data, such as files and images. It can also be used to relay messages to other Winlink users, making it a powerful tool for emergency response teams working in remote or isolated areas.

Overall, Winlink is an advanced digital mode that offers ham radio operators a powerful tool for communication in emergency situations. Its ability to work on low power, flexible transmission modes, and support for email and other data types make it an essential tool for emergency response teams and other organizations that require reliable communication during a crisis.

Satellite communications

Satellite communications can be an important tool for ham radio operators, especially during emergency situations when terrestrial communication systems may be down or overloaded. By using satellites in low Earth orbit or geostationary orbit, ham radio operators can communicate over long distances and across borders.

Satellite communication systems can offer several advantages over terrestrial systems. For example, they can provide a higher degree of reliability, security, and privacy compared to other forms of communication. They can also support a wider range of frequencies and modes, including voice, data, and video.

To use satellite communications, ham radio operators must have specialized equipment, including antennas, transceivers, and tracking systems. They must also have knowledge of orbital mechanics and the technical aspects of satellite communication systems.

Satellite communication systems can be used for a variety of purposes, including providing communications support during disaster response efforts, supporting remote operations in remote locations, and providing communications links for maritime and aviation operators.

To effectively use satellite communication systems, ham radio operators must have a clear understanding of the capabilities and limitations of the system, as well as the protocols and procedures used for satellite communications. They must also be familiar with the legal and regulatory requirements for using satellite systems, as well as the technical standards for satellite communications.

Overall, satellite communications can be a valuable tool for ham radio operators in emergency situations and in other contexts. By understanding the technical aspects of satellite communication systems and the protocols and procedures for using them, ham radio operators can effectively leverage satellite technology to support their communication efforts.

Amateur Radio Satellites

Amateur radio satellites can provide a unique and exciting opportunity for communication with other operators around the world. These satellites orbit the Earth and can be accessed with a variety of equipment and techniques. Here are some steps for using amateur radio satellites:

1. Research available satellites: Before attempting to use amateur radio satellites, it is important to research the available satellites and their characteristics. This infor-

mation can be found online or in amateur radio guides and manuals.

2. Obtain suitable equipment: To communicate with amateur radio satellites, you will need a suitable radio and antenna system. You may also need additional equipment such as preamps, power amplifiers, and tracking systems depending on your setup.
3. Learn about satellite tracking: Tracking satellites can be a complex process, and it is important to understand the basics of satellite tracking and the tools needed to do so. There are many resources available online and in amateur radio publications to help you learn more.
4. Determine pass times: Amateur radio satellites orbit the Earth on specific paths and schedules, and it is important to know when they will be passing over your location. There are many online tools and apps that can help you determine the pass times for specific satellites.
5. Prepare your station: Once you have determined the pass times for a satellite, you will need to prepare your station for communication. This may include setting up your antenna system, tuning your radio, and preparing any additional equipment you may need.
6. Communicate: When the satellite passes overhead, you can attempt to communicate with other operators. Be sure to follow proper communication protocols and procedures, and keep track of any contacts you make.
7. Monitor for future passes: Many amateur radio satellites make multiple passes over a specific location, so it is important to monitor for future passes and continue to communicate with other operators.

Note: It is important to follow all applicable laws and regulations when using amateur radio satellites, and to respect the frequencies and schedules used by other operators.

Commercial Satellites

Commercial satellites can also provide emergency communication capabilities in situations where traditional communication methods are unavailable or unreliable. These satellites can be used to provide high-speed data, voice, and video communication services to first responders and emergency management officials during natural disasters, terrorist attacks, and other emergency situations.

To use commercial satellites for emergency communication, you will need to:

1. Identify the appropriate satellite service provider: Several satellite service providers offer emergency communication services, including Inmarsat, Iridium, and Globalstar.
2. Determine the type of equipment needed: The type of equipment needed will depend on the type of satellite service provider and the specific services required. Some common equipment includes satellite phones, modems, and handheld devices.
3. Obtain the necessary licenses: Some countries may require licenses to operate satellite communication equipment. Check with your local regulatory agency to determine if you need a license.
4. Familiarize yourself with the equipment: Before using the equipment in an emergency situation, it is important to be-

come familiar with its operation and capabilities. Practice using the equipment during non-emergency situations to ensure that you know how to use it effectively.

5. Develop communication protocols: Develop communication protocols and procedures for using the equipment during an emergency. This should include procedures for initiating and terminating communication, as well as guidelines for transmitting and receiving information.
6. Test the equipment: Test the equipment to ensure that it is functioning properly before an emergency situation arises. This includes testing the equipment's ability to establish and maintain communication with the satellite.
7. Use the equipment during emergency situations: During an emergency situation, use the satellite communication equipment to transmit and receive critical information. Remember to follow established communication protocols and procedures to ensure effective communication.

Commercial satellites can provide a valuable backup communication option during emergency situations. By understanding how to use this technology and following best practices for emergency communication, you can help ensure effective communication and response during emergencies.

Mesh Networking

Mesh networking is a type of communication network in which each node or device is capable of communicating with multiple other nodes. This type of network is typically decentralized, meaning there is no central point of control or administration. Mesh networks are often used for emergency communication, as they can be quickly set up and provide a resilient and reliable means of communication.

To set up a mesh network, you will need several wireless mesh nodes, which act as the backbone of the network. These nodes can be set up in strategic locations to ensure maximum coverage, and can be connected to other nodes via wired or wireless connections. Each node is capable of forwarding messages to other nodes, so the network can continue to function even if some nodes are damaged or go offline.

In addition to mesh nodes, you will need software that can manage the network, such as a mesh networking protocol. There are several open source mesh networking protocols available, such as BATMAN-ADV, OLSR, and Babel, which can be customized to meet your specific needs.

To use a mesh network for emergency communication, you will need to ensure that each node has a power source, such as a battery or generator, and a device for sending and receiving messages, such as a radio or smartphone. You will also need to establish communication protocols and procedures, such as assigning channels and frequencies, setting up message forwarding rules, and defining emergency message formats.

One advantage of using a mesh network for emergency communication is that it can be set up quickly and easily, even in remote or hard-to-reach areas. Mesh networks can also be easily expanded or modified as needed, and can provide a high degree of reliability and redundancy.

However, there are also some challenges associated with mesh networks, such as the need for sufficient power and bandwidth to support multiple nodes, and the potential for interference and congestion as more nodes are added to the network. It is also important to ensure that the network is secure and that only authorized users have access to it.

Use in Emergencies

Mesh networks can be a valuable tool for emergency communication because they are decentralized, meaning that they can operate without relying on a centralized infrastructure such as the internet. In a mesh network, nodes communicate with each other directly, allowing for messages to be relayed over multiple paths to reach their destination. This makes mesh networks resilient to failures and disruptions in the network, making them a reliable communication option in emergency situations.

Here are some steps for using mesh networks for emergency communications:

1. Set up a mesh network: To set up a mesh network, you will need nodes (devices that can communicate with each other), mesh networking software, and a power source (such as batteries or generators). You can purchase pre-built mesh networking equipment or build your own using off-the-shelf hardware and open-source software.
2. Train volunteers: Once your mesh network is set up, train volunteers on how to use the network for emergency communications. Make sure they understand how to access the network, how to send and receive messages, and how to troubleshoot any issues that may arise.
3. Establish communication protocols: To ensure efficient communication, establish communication protocols that are compatible with the mesh network software you are

using. This could include using specific channels or frequencies for certain types of messages, or using specific codes or abbreviations to indicate the urgency or priority of a message.

4. Test the network: Before relying on the mesh network for emergency communications, test it thoroughly to ensure that it is working properly. This could include running simulations of different emergency scenarios to identify any weaknesses in the network.
5. Work with local authorities: Coordinate with local emergency response officials to ensure that your mesh network is integrated into their emergency communication plans. This could include sharing contact information, providing regular updates on the status of the mesh network, and participating in emergency response drills and exercises.
6. Continuously improve: As you use the mesh network for emergency communications, continue to monitor its performance and make improvements where necessary. This could include upgrading equipment, expanding the network coverage area, or updating communication protocols to better meet the needs of emergency responders.

By using mesh networks for emergency communications, you can ensure that communication remains available and reliable even in the face of infrastructure failures or other disruptions. With proper planning, training, and coordination with local authorities, mesh networks can play a critical role in emergency response efforts.

9

The Future of Emergency Communications

prototype of R18 octocopter developed by Aerorozvidka

The future of emergency communications is constantly evolving, with new technologies and techniques being developed to improve communication during emergency situations. Here are some potential developments that could shape the future of emergency communications:

1. Use of Drones: Drones can be used to quickly survey and assess disaster zones, and can also be used to deliver emergency supplies and equipment to inaccessible areas. In the future, drones may be used to provide communication relays or even provide temporary cellular service in disaster areas.
2. Artificial Intelligence (AI): AI can be used to quickly analyze large amounts of data to identify patterns and provide insights that can aid emergency response efforts. In the future, AI could be used to analyze social media and other online sources to quickly identify areas that need emergency assistance.
3. Internet of Things (IoT): IoT devices, such as sensors and cameras, can be used to monitor conditions in disaster areas and provide real-time information to emergency responders. In the future, the widespread adoption of IoT devices could greatly improve situational awareness and help emergency responders make better decisions.
4. 5G Networks: The rollout of 5G networks could greatly improve communication during emergency situations, providing faster and more reliable connections. In addition, the low latency of 5G networks could make it possible to remotely control emergency response robots and drones.
5. Virtual Reality (VR): VR technology can be used to provide

immersive training for emergency responders and to simulate disaster scenarios for planning purposes. In the future, VR could be used to provide virtual tours of disaster zones to aid in situational awareness.

6. Quantum Communications: Quantum communications is a technology that uses the principles of quantum mechanics to transmit information securely over long distances. In the future, quantum communications could be used to provide secure, high-speed communication during emergency situations.

Overall, the future of emergency communications is likely to be shaped by a combination of new technologies and innovations. By staying up-to-date with the latest developments in the field, emergency responders and ham radio operators can be better prepared to communicate during times of crisis.

Use of Drones

The use of drones for emergency communications is an emerging technology that has the potential to revolutionize emergency response efforts. Drones can provide situational awareness, deliver supplies, and provide communication support in disaster areas where traditional infrastructure has been damaged or destroyed. This section will cover the current and future use of drones in emergency communications.

Drones equipped with cameras and other sensors can provide real-time situational awareness to emergency responders. They can be used to survey areas that are difficult to access or are unsafe for first responders, such as collapsed buildings or flooded areas. By providing high-resolution images and video feeds, drones can help emergency responders to better understand the extent of the damage and to plan their response efforts accordingly.

Drones can also be used to deliver supplies and equipment to disaster areas. They can be used to transport medical supplies, food, and other critical items to areas that are difficult to access due to damaged infrastructure or hazardous conditions. Drones can also be used to transport communication equipment, such as radio repeaters or cellular base stations, to disaster areas where traditional infrastructure has been damaged or destroyed.

In addition to providing situational awareness and delivering supplies, drones can also be used to provide communication support in disaster areas. Drones equipped with communication equipment, such as Wi-Fi or cellular base stations, can be used to establish temporary communication networks in disaster areas. This can help emergency responders to coordinate their efforts and communicate with each other more effectively.

The future of emergency communications with drones is promising, with ongoing research and development focused on improving the capabilities of drones for emergency response efforts. Future drones may have increased endurance, longer range, and improved communication capabilities. They may also be equipped with advanced sensors and artificial intelligence capabilities to better detect and respond to emergency situations.

Overall, the use of drones for emergency communications is an exciting and rapidly evolving technology. As the capabilities of drones continue to improve, they will play an increasingly important role in emergency response efforts, helping to save lives and minimize damage during crisis situations.

Artificial Intelligence (AI)

Artificial Intelligence (AI) has the potential to revolutionize emergency communications. AI technology can be used to process vast amounts of data in real-time, enabling emergency responders to make quick and informed decisions during crisis situations. AI algorithms can be trained to recognize patterns in data, identify anomalies, and predict potential problems before they occur. This can help emergency responders to better prepare for and respond to emergencies.

In addition, AI can be used to improve the accuracy of emergency alerts and warnings. Natural language processing (NLP) technology can be used to analyze text messages and social media posts in real-time to identify potential emergency situations. AI-powered chatbots can also be used to provide real-time information and assistance to individuals in emergency situations.

AI can also be used to enhance the capabilities of emergency communication systems. For example, AI-powered speech recognition technology can be used to transcribe voice messages and convert them into text. This can help emergency responders to quickly process and respond to emergency messages. AI can also be used to automatically identify and prioritize emergency messages based on their content and urgency.

In the future, AI could be used to create autonomous emergency response systems that can operate without human intervention. For example, drones equipped with AI technology could be used to assess damage and identify areas that need immediate attention. AI could also be used to develop self-driving emergency vehicles that can quickly transport supplies and equipment to emergency sites.

Overall, the potential of AI in emergency communications is vast. As the technology continues to evolve, it will likely play an increasingly important role in improving emergency response and communication capabilities.

Internet of Things (IoT)

The Internet of Things (IoT) is a term used to describe a network of physical devices, vehicles, buildings, and other items embedded with electronics, software, sensors, and connectivity which enable these objects to collect and exchange data. In the context of emergency communications, the IoT has the potential to improve situational awareness and response times.

For example, during a natural disaster, IoT devices such as weather sensors, water level sensors, and traffic cameras can be used to provide real-time information about the situation on the ground. This information can then be used to guide emergency response efforts and help responders make more informed decisions.

IoT devices can also be used to track the location of people during emergencies. For example, wearable devices such as smartwatches and fitness trackers can be used to monitor the location and movement of individuals, which can be especially useful in situations where people are missing or trapped.

Another potential application of IoT in emergency communications is the use of drones equipped with sensors and cameras. These drones can be used to provide real-time video feeds of disaster zones and other emergency situations, allowing responders to assess the situation and plan their response accordingly.

Overall, the IoT has the potential to greatly enhance emergency communications and response efforts. By providing real-time data and situational awareness, IoT devices can help responders make more informed decisions and improve their response times. However, the use of IoT in emergency communications also raises concerns about privacy and security, which must be addressed in order to ensure the safe and effective use of these technologies.

5G Networks

The emergence of 5G networks represents a significant development in the field of emergency communications. With faster speeds, lower latency, and greater capacity, 5G networks can enable a range of new capabilities for emergency responders and organizations. Here are some ways that 5G networks could impact emergency communications:

1. Enhanced situational awareness: 5G networks can support high-speed data transmission, which can enable the use of real-time video and other sensor data for situational awareness. This could include the use of drones, cameras, and other IoT devices to provide first responders with a better understanding of the emergency situation.
2. Improved communication between responders: With the increased capacity of 5G networks, emergency responders can communicate more effectively with each other. This can include voice and video communication, as well as the use of advanced applications for collaboration and data

sharing.

3. Faster response times: With the lower latency of 5G networks, emergency responders can access critical information and respond to emergencies more quickly. This can be especially important in situations where seconds can make a difference.
4. Remote monitoring and control: The low latency and high-speed data transmission of 5G networks can enable remote monitoring and control of emergency systems and devices. This could include the use of remote sensors to detect and respond to emergencies, as well as the use of remote control systems to manage response efforts.
5. Smart cities: With the advent of 5G networks, there is the potential to create smart cities that are more resilient to emergencies. This could include the use of sensors and other IoT devices to detect and respond to emergencies, as well as the use of advanced analytics to better understand and prepare for potential emergencies.

Overall, the emergence of 5G networks represents a significant opportunity for the field of emergency communications. By leveraging the increased capacity, speed, and low latency of 5G networks, emergency responders and organizations can improve situational awareness, communication, and response times. As 5G networks continue to evolve, it will be important to explore new applications and capabilities for emergency communications.

Virtual Reality (VR)

Virtual Reality (VR) is an emerging technology that could potentially play a role in emergency communications. VR involves the use of headsets or other devices that allow individuals to experience a simulated, three-dimensional environment. In the context of emergency communications, VR could be used to provide immersive training experiences for first responders and emergency management officials.

One potential use of VR in emergency communications is for disaster preparedness and response training. VR could allow individuals to experience simulated emergency scenarios in a safe and controlled environment, helping them to develop the skills and knowledge needed to respond effectively in real-world emergency situations.

Another potential use of VR in emergency communications is for situational awareness. By using VR to visualize data from sensors and other sources, emergency management officials and first responders could gain a better understanding of the situation on the ground, allowing them to make more informed decisions.

VR could also be used to provide remote assistance during emergency situations. For example, medical professionals could use VR to provide remote guidance to first responders in the field, helping them to make better decisions and provide more effective care.

However, there are also potential challenges to the use of VR in emergency communications. VR equipment can be expensive and may not be readily available in all locations. Additionally, the use of VR may require specialized training and expertise, which could limit its usefulness in some situations.

Overall, VR is an emerging technology with the potential to enhance emergency communications and response efforts. While there are challenges to its use, ongoing advancements in VR technology and increased availability could make it a valuable tool for emergency management and first responders in the future.

Quantum Communications

Quantum communications is a new area of research in the field of communication technology. It uses the principles of quantum mechanics to send and receive secure information over long distances. Traditional communication methods rely on the exchange of electronic signals, which can be intercepted and compromised. In contrast, quantum communication relies on the transmission of single photons, which cannot be intercepted without altering their state and therefore alerting the receiver.

Quantum communication offers several advantages over traditional communication methods, including:

1. Security: Quantum communication offers a level of security that cannot be achieved with traditional communication methods. Since single photons are used to transmit information, any attempt to intercept the signal will alter its state, making it impossible to decipher the information without detection.
2. Efficiency: Quantum communication can transmit information over long distances with minimal loss of data or signal quality. This makes it an attractive option for

applications where high-speed, reliable communication is essential.

3. Scalability: Quantum communication can be easily scaled up to accommodate larger networks, making it an ideal solution for applications where multiple nodes need to communicate with each other.
4. Resilience: Quantum communication is highly resistant to interference from external sources, such as electromagnetic radiation or other signals. This makes it an ideal solution for applications where reliability and resilience are essential.

Despite its many advantages, quantum communication is still in the early stages of development, and there are many challenges that need to be overcome before it can become a practical solution for emergency communication. These challenges include the need for specialized equipment and infrastructure, as well as the development of new protocols and standards for quantum communication. However, as research in this area continues, it is likely that quantum communication will play an increasingly important role in the future of emergency communication.

10

Case Studies

Hurricane Katrina, the Nepal earthquake, and the California wildfires were devastating natural disasters that affected thousands of people. During these crises, amateur radio operators played a crucial role in providing emergency communication and support to those affected.

In the aftermath of Hurricane Katrina, which struck the Gulf Coast of the United States in 2005, amateur radio operators were able to provide vital communication links when traditional means of communication failed. They set up temporary repeater stations, established communication links between emergency response agencies, and provided communication support for medical and rescue teams. Ham radio operators also provided support for public service and humanitarian organizations, which helped to coordinate the response effort and provide critical aid to those affected by the disaster.

In 2015, a massive earthquake struck Nepal, killing thousands and leaving many more injured and homeless. In the aftermath of the disaster, amateur radio operators were again able to provide critical communication support to those affected.

They helped to establish communication links between relief organizations and remote areas, provided support for medical teams and first responders, and assisted in coordinating rescue efforts. Ham radio operators also worked closely with local government agencies to provide real-time information on the situation on the ground, which helped to guide the response effort.

The California wildfires of 2018 were one of the most devastating wildfire seasons in the state's history, causing widespread destruction and forcing thousands to evacuate their homes. In the face of this disaster, amateur radio operators again played a critical role in providing emergency communication and support. They helped to coordinate evacuations, set up communication links between emergency response agencies, and provided real-time information on the spread of the wildfires. Ham radio operators also assisted in coordinating relief efforts and providing support for medical and rescue teams working in the affected areas.

Overall, the role of amateur radio operators in providing emergency communication and support during natural disasters cannot be overstated. Their skills, experience, and equipment allow them to provide critical communication links when traditional means of communication fail, and they play a vital role in coordinating response efforts and providing support to those affected by disasters.

Hurricane Katrina

Hurricane Katrina was a devastating Category 5 hurricane that hit the Gulf Coast of the United States in August 2005. It was one of the deadliest and costliest natural disasters in

U.S. history. The hurricane caused widespread destruction in Louisiana, Mississippi, and Alabama, and its aftermath resulted in significant social and economic impacts.

The hurricane made landfall near New Orleans, Louisiana, and caused extensive flooding due to the failure of the city's levee system. More than 1,800 people died as a result of the storm, and thousands were injured or left homeless. The hurricane also caused billions of dollars in property damage and economic losses.

The disaster highlighted the importance of effective emergency communication and response. The breakdown of communication systems during the hurricane made it difficult for emergency responders to coordinate their efforts, and many people were left stranded or without basic necessities such as food, water, and medical supplies.

The use of ham radio operators played a critical role in emergency communication during Hurricane Katrina. Ham radio operators were able to provide real-time communication in areas where other communication systems had failed, allowing emergency responders to coordinate their efforts and provide assistance to those in need. Ham radio operators were also able to relay important information about the storm's impact and assist in search and rescue efforts.

The hurricane prompted significant improvements in emergency communication and response systems. It led to the establishment of new protocols for emergency communication, the development of new technologies for disaster response, and improvements in emergency preparedness and training. The lessons learned from Hurricane Katrina have helped to improve the effectiveness of emergency communication and response efforts in subsequent disasters.

Overall, Hurricane Katrina was a tragic reminder of the importance of effective emergency communication and response. The use of ham radio operators played a critical role in providing communication during the disaster, and the lessons learned have helped to improve emergency preparedness and response efforts in subsequent disasters.

Nepal Earthquake

In 2015, Nepal was hit by a devastating 7.8 magnitude earthquake, causing widespread destruction and loss of life. The earthquake also triggered a massive avalanche on Mount Everest, further adding to the chaos and destruction. In the aftermath of the earthquake, communication infrastructure was severely damaged, making it difficult for emergency responders to coordinate their efforts and for individuals to get in touch with their loved ones.

Amateur radio operators played a critical role in providing emergency communication support in Nepal. Ham radio operators from around the world joined forces to establish communication links with the affected areas, providing a lifeline for emergency responders and individuals trying to get in touch with their loved ones. With the help of ham radio operators, critical medical supplies and other essential resources were quickly transported to the affected areas, helping to save lives and alleviate suffering.

Despite the challenges posed by the difficult terrain and damaged infrastructure, ham radio operators were able to establish communication links with remote villages and other hard-to-reach areas. These communication links were essential in helping emergency responders coordinate their efforts and ensure that critical resources were distributed where they were needed most.

Through their dedication and expertise, ham radio operators were able to make a significant contribution to the relief efforts in Nepal, highlighting the critical role that amateur radio can play in emergency situations.

California Wildfires

In October 2017, a series of wildfires broke out in Northern California, leading to the destruction of thousands of homes and businesses and the displacement of tens of thousands of people. Ham radio operators played a critical role in providing emergency communication during the wildfires, as the destruction of communication infrastructure made it difficult for first responders to coordinate their efforts.

Ham radio operators set up communication centers in areas affected by the wildfires, using portable equipment and generators to power their stations. They also worked closely with local authorities to coordinate communication efforts and provide situational awareness to first responders. Ham radio operators were able to provide real-time information about the spread of the fires, the status of evacuation efforts, and the location of emergency shelters.

One notable example of the effectiveness of ham radio communication during the California wildfires was the case of a family trapped in their home as the fire approached. They were able to use their ham radio to contact a nearby repeater, which relayed their distress call to a ham radio operator at an emergency communication center. The ham radio operator was then able to coordinate with local authorities to rescue the family before their home was consumed by the fire.

Overall, the California wildfires demonstrated the critical role that ham radio operators can play in emergency communication, particularly when traditional communication infrastructure is disrupted. Ham radio operators were able to provide vital communication support to first responders and affected communities, helping to save lives and minimize damage during the crisis.

11

If you've read my book,

If you've read my book, I would be grateful if you could take a moment to leave an honest review on Amazon. Your review will not only help other readers make an informed decision but also provide valuable feedback to me as an author. Thank you for taking the time to share your thoughts!

Glossary

1. Amateur Radio: A form of communication where individuals use designated radio frequencies for non-commercial purposes.
2. Antenna: A device used to transmit and receive radio signals.
3. APRS (Automatic Packet Reporting System): A digital communication system used to send location and other data over amateur radio.
4. Band: A range of frequencies allocated for a particular purpose or type of transmission.
5. Call Sign: A unique identification assigned to an amateur radio operator.
6. CQ: A general call to any station.
7. DX: Distance communication between stations.
8. Echolink: A computer program that allows amateur radio operators to communicate over the internet.
9. Emergency Communications: Communications used during an emergency or disaster.
10. FCC (Federal Communications Commission): A government agency responsible for regulating communication in the United States.
11. Field Day: An annual event where amateur radio operators set up temporary stations to test their communication

skills.

12. Frequency: The number of cycles per second of a radio wave.
13. HAM: A term used to refer to an amateur radio operator.
14. IARU (International Amateur Radio Union): An organization that represents amateur radio operators worldwide.
15. J-Pole: A type of antenna often used by amateur radio operators.
16. Kilohertz (kHz): A unit of measurement used to describe radio frequencies.
17. Ladder Line: A type of transmission line often used with dipole antennas.
18. License: An official document that permits an individual to operate an amateur radio station.
19. Linear Amplifier: An electronic device used to increase the power of a radio signal.
20. Morse Code: A system of communication using a series of dots and dashes.
21. Net: A group of stations that communicate on a regular basis.
22. Packet Radio: A digital mode of communication used by amateur radio operators.
23. Portable: A term used to describe equipment that can be easily transported
24. QSL Card: A postcard used to confirm a contact between two amateur radio operators.
25. Repeater: A device used to receive a signal and retransmit it at a higher power.
26. RF (Radio Frequency): The portion of the electromagnetic spectrum used for radio communication.
27. RST System: A system used to report signal quality in

amateur radio communication.

28. Shortwave: Frequencies used for long-distance communication.
29. SWR (Standing Wave Ratio): A measure of the efficiency of an antenna.
30. Technician Class License: The entry-level license for amateur radio operators in the United States.
31. Transceiver: A device that can transmit and receive radio signals.
32. UHF (Ultra High Frequency): Frequencies between 300 MHz and 3 GHz.
33. VHF (Very High Frequency): Frequencies between 30 MHz and 300 MHz.
34. VOIP (Voice over Internet Protocol): A method of transmitting voice communication over the internet.
35. Watt: A unit of measurement used to describe the power output of a radio signal.
36. XYL: A term used to refer to a ham radio operator's wife.
37. Yagi: A directional antenna often used by amateur radio operators.
38. ZULU Time: The time used in aviation and military operations, also known as Coordinated Universal Time (UTC).
39. APRS-IS (Automatic Packet Reporting System Internet Service): The global network that connects APRS stations.
40. Digital Signal Processing: The use of computer algorithms to manipulate radio signals.
41. Ground Plane: An antenna that uses a flat plane of metal as a reflector to increase the signal.
42. Propagation: The way radio waves travel through the atmosphere.
43. Antenna Tuner: a device used to match the impedance of

an antenna system to the transmitter's output, ensuring maximum power transfer between the two.

44. QRP: A ham radio operating mode where the operator uses low power, typically less than 5 watts, to communicate over short distances.
45. RACES: The Radio Amateur Civil Emergency Service is a volunteer organization of amateur radio operators who provide communication support during emergencies and disasters.
46. Repeater: A device that receives a signal on one frequency and retransmits it on another frequency to increase the range of communication.
47. RF: Short for radio frequency, this term refers to the frequency range used for radio communication.
48. Rig: Slang term for a ham radio transceiver.
49. SWR: Standing Wave Ratio, a measure of the efficiency of an antenna system.
50. VOIP: Voice over Internet Protocol, a communication technology that allows voice communication over the internet.

Made in United States
Orlando, FL
28 April 2023